Heat Transfer Examples

Practical Problems Solved

by D. James Benton

Preface

This is a collection of increasingly complex solved problems involving heat transfer, which must also call upon the principles of thermodynamics and material properties. A working knowledge of differential equations is also essential to understanding this subject. We will cover thermal aspects and also pressure drop but neither mechanical concerns, such as structural, nor chemical ones, such as corrosion. While these latter considerations are important in any design, they must be addressed elsewhere. All examples are worked in SI units with English units in parentheses. A practicing engineer should be comfortable working with any units.

All of the examples contained in this book,
(as well as a lot of free programs) are available at...

https://www.dudleybenton.altervista.org/software/index.html

Figure 1. Typical Air-Cooled Finned-Tube Heat Exchanger

Table of Contents

Figure 2. Typical Plate Heat Exchanger

Example 1. One-Dimensional Steady-State Conduction

The first example we consider is the simplest (computationally): one-dimensional steady-state conduction. This figure illustrates the problem.

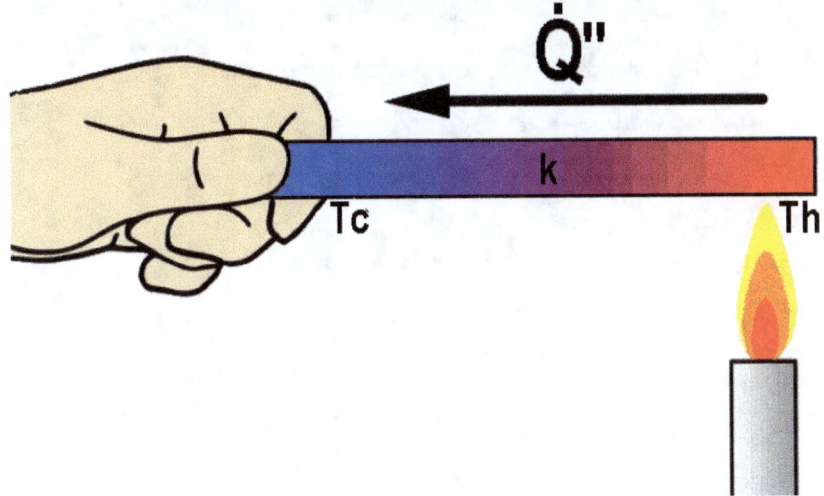

Figure 3. Heated Rod

For this example we will ignore two- and three-dimensional effects and presume the temperature of the rod only varies along the length. We will also ignore transients. Obviously, it would take a finite amount of time for the rod to heat up and the temperature of the end we are holding will change over time as well. The rate of heat transfer per unit area (heat flux) in this case is given by Fourier's Law:[1]

$$\dot{Q}'' = \frac{\partial Q}{A \partial t} = k\frac{\partial T}{\partial x} \approx k\frac{\Delta T}{\Delta x} \approx \frac{k}{L}\left(T_h - T_c\right) \tag{1.1}$$

where Q is the heat, T is the temperature, k is the thermal conductivity, and L is the length of the rod, say 20 cm (8 in). If the rod were steel, k would be about 40 W/m/°C (23 BTU/hr/ft/°F). The temperatures might be 250°C (482°F) and 25°C (77°F), respectively. The heat flux is then:

$$\dot{Q}'' = \frac{\left(40\,\dfrac{W}{m°C}\right)}{(0.2m)}\left(250°C - 25°C\right) = 45{,}000\,\frac{W}{m^2} \tag{1.2}$$

or 45 kW/m² (14,265 BTU/hr/ft²). See spreadsheet examples.xls for details (shown below).

---

[1] Jean-Baptiste Joseph Fourier (1768–1830) French mathematician and physicist

Example 1. Calculations

	A	B	C	D	E	F
1	**1D Steady-State Conduction**					
2	Sample Calculations		SI Units		English Units	
3	INPUTS	symbol	units	value	units	value
4	rod length	L	m	0.20	ft	0.656
5	conductivity	k	W/m/°C	40	BTU/hr/ft/°F	23.1
6	hot temperature	Th	°C	250	°F	482
7	cold temperature	Tc	°C	25	°F	77
8	SOLUTION					
9	heat flux	Q	kW/m²	45	BTU/hr/ft²	14,274
10						
11			user inputs in blue			
12			calculations in orange			

Example 2. One-Dimensional Steady-State Convection

We next consider a loaf of bread fresh out of the oven.

Figure 4. Warm Loaf of Bread

The surface of the bread is at 180°C (356°F) and the room is at 25°C (77°F). A typical heat transfer coefficient, h, for natural convection in air is about 1 W/m²/°C (0.176 BTU/hr/ft²/°F). The heat flux in this case is given by:

$$\dot{Q}'' = h(T_h - T_c) = \left(1\frac{W}{m^2\,{}^\circ C}\right)(180°C - 25°C) = 155\frac{W}{m^2} \tag{2.1}$$

or 0.155 kW/m² (49 BTU/hr/ft²). See spreadsheet examples.xls for details. These first few problems primarily introduce variables, equations, properties, and units. The Engineering Toolbox is a good online source of thermal and transport properties:

https://www.engineeringtoolbox.com/

3

Example 2. Calculations

	A	B	C	D	E	F
1	**1D Steady-State Convection**					
2	Sample Calculations		SI Units		English Units	
3	INPUTS	symbol	units	value	units	value
4	ht. tr.coef.	h	W/m²/°C	1	BTU/hr/f²t/°F	0.176
5	hot temperature	Th	°C	180	°F	356
6	cold temperature	Tc	°C	25	°F	77
7	SOLUTION					
8	heat flux	Q	W/m²	155	BTU/hr/ft²	49.1
9						
10			user inputs in blue			
11			calculations in orange			

4

Example 3: One-Dimensional Steady-State Radiation

The surface of the Sun is effectively 5778°K (10,400°R). Radiative heat transfer from the Sun is radially outward in all directions and so one-dimensional for our purposes. The heat flux is given by the Stefan-Boltzmann[2] Law and proportional to the difference in the fourth power of the temperatures:

$$\dot{Q}'' = \varepsilon\sigma\left(T_h^4 - T_c^4\right)$$

(3.1)

where ε is the emissivity (unitless) and σ is the Stefan-Boltzmann[3] constant ($\sigma = 5.670374419\times10^{-8}$ W/m²/K⁴).

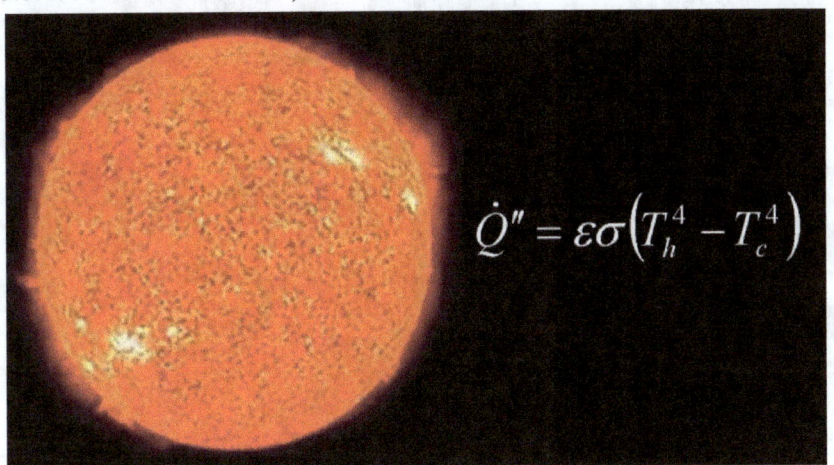

Figure 5. Solar Radiative Heat Transfer

The effective emissivity of the Sun is about 0.99, which is indicative of black-body radiation and is tied to the effective surface temperature in this case. The effective temperature of space is about 2.73K. The heat flux is then calculated:

$$\dot{Q}'' = 0.99\left(5.67\times10^{-8}\,\frac{\text{W}}{\text{m}^2\text{K}^4}\right)\left(5778^4 - 2.73^4\right)\text{K}^4$$

$$= 6.26\times10^7\,\frac{W}{m^2}$$

(3.2)

or 6.26×10^4 kW/m² (19.8×10^6 BTU/hr/ft²). See spreadsheet examples.xls for details. Additional information on radiation heat transfer can be found at the Engineering Toolbox plus there is an article on Wikipedia:

https://en.wikipedia.org/wiki/Thermal_radiation_-_Interchange_of_energy

---

[2] Josef Stefan (1835–1893) Austrian physicist, mathematician, and poet
[3] Ludwig Eduard Boltzmann (1844–1906) Austrian physicist and philosopher

Example 3. Calculations

	A	B	C	D	E	F
1	**1D Steady-State Radiation**					
2	Sample Calculations		SI Units		English Units	
3	INPUTS	symbol	units	value	units	value
4	emissivity	ε	-	0.99	-	0.990
5	S-B constant	σ	W/m²/°K⁴	5.67E-08	BTU/hr/ft²/°R⁴	1.71E-09
6	hot temperature	Th	°K	5778	°R	10400
7	cold temperature	Tc	°K	2.73	°R	5
8	SOLUTION					
9	heat flux	Q	kW/m²	6.26E+07	BTU/hr/ft²	1.98E+07
10						
11			user inputs in blue			
12			calculations in orange			

Example 4: Forced Convection Inside A Pipe

We next consider fully-developed turbulent forced convection inside a round pipe. We will consider laminar flow, developing flow, entrance and exit losses, and fittings in a subsequent chapter. Thermal resistance of the pipe wall and what happens on the outside of the pipe will also be covered in a subsequent chapter.

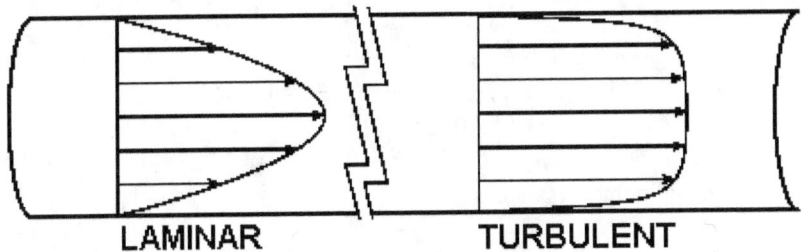

LAMINAR   TURBULENT

Figure 6. Flow Inside A Pipe

The heat flux in this case is the same as Equation 2.1 only we now consider the calculation of the heat transfer coefficient, h. For fully-developed turbulent flow in a pipe, we use the correlation of Dittus and Boelter[4]:

$$Nu = 0.023\, Re^{\frac{4}{5}}\, Pr^{n} \tag{4.1}$$

where n=0.4 for heating and n=0.3 for cooling. Nu is the Nusselt number:

$$Nu = \frac{hd}{k} \tag{4.2}$$

where d is the diameter and k is the thermal conductivity of the fluid. Re is the Reynolds number:

$$Re = \frac{\rho V d}{\mu} \tag{4.3}$$

where ρ is the fluid density, V is the mean fluid velocity, and μ is the dynamic viscosity of the fluid. There are corrections for variable properties, which we will discuss later. Pr is the Prandtl number:

$$Pr = \frac{\mu C}{k} \tag{4.4}$$

where C is the specific heat of the fluid.

---

[4] Dittus, P. W. and L. M. Boelter, L.M., University of California Publications in Engineering, Vol. 1, No. 13, pp. 443-461 1930 (reprinted in *International Communications in Heat and Mass Transfer*, Vol. 12, pp. 3-22, 1985).

We first need dimensions and properties. The inside diameter of the pipe is 5 cm (1.97 in). The fluid is water and the mean velocity is 2 m/s (6.56 ft/sec). The mean temperature of the water is 25°C (77°F). The inside wall temperature of the pipe is 75°C (167°F). The density is 997 kg/m³ (62.26 lbm/ft³). The dynamic viscosity is 0.9 cP (2.18 lbm/ft/hr). The thermal conductivity is 0.6 W/m/°C (0.35 BTU/hr/ft/°F). The specific heat is 4.18 kJ/kg/°C (1 BTU/lbm/°F). The Reynolds number is then:

$$\mathrm{Re} = \frac{\left(997\frac{kg}{m^3}\right)\left(2\frac{m}{s}\right)(0.05m)}{\left(0.0009\frac{kg}{ms}\right)} = 110,800 \tag{4.5}$$

The Prandtl number is then:

$$\mathrm{Pr} = \frac{\left(0.0009\frac{kg}{ms}\right)\left(4180\frac{J}{kg°C}\right)}{\left(0.6\frac{W}{m°C}\right)} = 6.27 \tag{4.6}$$

The Nusselt number is then:

$$Nu = 0.023(110,800)^{\frac{4}{5}}(6.27)^{0.4} = 520 \tag{4.7}$$

The heat transfer coefficient is then:

$$h = \frac{kNu}{d} = \frac{\left(0.6\frac{W}{m°C}\right)520}{(0.05m)} = 6243\frac{W}{m^2°C} \tag{4.8}$$

or 6.24 kW/m²/°C (1100 BTU/hr/ft²/°F). The heat flux is then:

$$\dot{Q}'' = h(T_h - T_c) = \left(6.24\frac{kW}{m^2°C}\right)(75°C - 25°C) = 312\frac{kW}{m^2} \tag{4.9}$$

or 98,900 BTU/hr/ft². See spreadsheet examples.xls for details.

Example 4. Calculations

	A	B	C	D	E	F
1	**Forced Convection inside a Pipe**					
2	Sample Calculations		SI Units		English Units	
3	INPUTS	symbol	units	value	units	value
4	tube inside dia.	d	m	0.05	ft	0.164
5	mean velocity	V	m/s	2	ft/sec	6.56
6	density	ρ	kg/m³	997	lbm/ft³	62.2
7	viscosity	μ	cP	0.90	lbm/ft/hr	2.18
8	conductivity	k	W/m/°C	0.60	BTU/hr/ft/°F	0.347
9	specific heat	C	kJ/kg/°C	4.18	BTU/lbm/°F	0.998
10	tube wall temp.	Th	°C	75	°F	167
11	mean fluid temp.	Tc	°C	25	°F	77
12	SOLUTION					
13	Reynolds Number	Re	-	1.11E+05	-	1.11E+05
14	Prandtl Number	Pr	-	6.27	-	6.27
15	Nusselt Number	Nu	-	520	-	520
16	ht. tr. coef.	h	W/m²/°C	6243	BTU/hr/f²t/°F	1100
17	heat flux	Q	kW/m²	3.12E+02	BTU/hr/ft²	9.90E+04
18						
19			user inputs in blue			
20			calculations in orange			

9

Example 5: Condensation Outside A Tube

We next consider the condensation of steam on the outside of tubes.

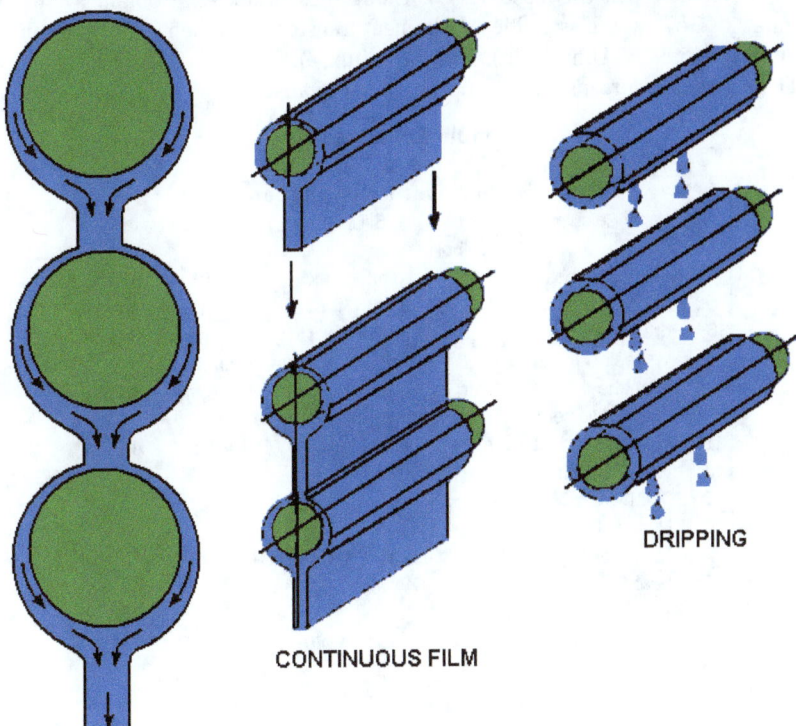

CONTINUOUS FILM

DRIPPING

Figure 7. Condensation on a Tube

The original experimental work on this type of condensation and associated formula for calculating the heat transfer coefficient is due to Nusselt.[5] This formula may be found in any heat transfer text as well as readily online.

$$Nu = 0.728 \left[\frac{\rho_f (\rho_f - \rho_g) h_{fg} g d^3}{k \mu \Delta T} \right]^{\frac{1}{2}}$$

(5.1)

where ρ_f and ρ_g are the densities of the saturated liquid and saturated vapor steam, respectively (995 and 0.0339 kg/m³); h_{fg} is the latent heat of condensation (2425 kJ/kg); g is the acceleration of gravity (9.8 m/s²); d is the outside diameter of the tubes (25 mm); k is the thermal conductivity of the saturated liquid (0.62 W/m/°C); μ is the dynamic viscosity of the saturated liquid (0.75 cP); and ΔT is

---

[5] Nusselt, W., "Die Oberflächenkondensation des Wasserdampfes [The Surface Condensation of Water Vapor]," VDI [Association of German Engineers], 1916.

the temperature difference across the condensing film of water (generally taken to be the saturation temperature of the surrounding steam minus the outside tube wall temperature, in this case 12°C). Using these values we obtain a Nusselt number of 367, which is unitless. The heat transfer coefficient, h, is then 9074 W/m²/°C (1599 BTU/hr/ft²/°F). The heat flux, q, is then 109 kW/m² (34,533 BTU/hr/ft²). See spreadsheet examples.xls for details.

Example 5. Calculations

	A	B	C	D	E	F
1	Condensation on a Tube					
2	Sample Calculations		SI Units		English Units	
3	INPUTS	symbol	units	value	units	value
4	sat. lqd. density	ρ_Y	kg/m³	995	lbm/ft³	62.12
5	sat. vap. density	ρ_g	kg/m³	0.0339	lbm/ft³	0.00211
6	gravity	g	m/s²	9.76	ft/sec²	32.018
7	viscosity	μ	cP	0.76	lbm/ft/hr	1.85
8	latent heat	hfg	kJ/kg	2425	BTU/lbm	1043
9	outside diameter	d	mm	25	in	0.984
10	thermal conduct.	k	W/m/°C	0.618	BTU/hr/ft/°F	0.357
11	temperature diff.	ΔT	°C	12	°F	21.6
12	SOLUTION					
13	Nusselt Number	Nu	-	367	-	367
14	ht. tr. coef.	h	W/m²/°C	9074	BTU/hr/ft²/°F	1599
15	heat flux	q	kW/m²	109		34,533
16						
17	user inputs in blue					
18	calculations in orange					

Example 6. Nucleate Boiling

Perhaps the simplest form of boiling is nucleate pool. Flow boiling has many more complexities, as does subcooled, incipient, superheated, and film boiling. Rohsenow developed the first empirical correlation for nucleate boiling, which is still widely used.[6] It can be expressed in a variety of ways, including:

$$\frac{\dot{Q}}{A} = \left(\frac{\mu h_{fg}}{Pr^{3n}}\right)\left[\frac{g(\rho_f - \rho_g)}{\sigma}\right]^{\frac{1}{2}}\left[\frac{C\Delta T}{h_{fg}c}\right]^3 \tag{6.1}$$

where μ is the dynamic viscosity of the liquid; h_{fg} is the latent heat of vaporization; Pr is the Prandtl number of the liquid; g is the surface tension of the liquid; ρ_f and ρ_g are the densities of the saturated liquid and saturated vapor steam, respectively; σ is the surface tension; C is the (constant pressure) specific heat of the liquid; ΔT is the excess temperature (hot surface temperature minus temperature of the liquid); and c is an empirical constant, which varies with the liquid and surface materials, roughness, and cleanliness. The exponent, n, on the Prandtl number in Equation 6.1 can vary from 1 to 2, depending on the liquid.

Perhaps the biggest challenge in using Rohsenow's correlation is selecting the right empirical factor, c, as there are endless combinations of liquids and surfaces materials plus a wide range of surface conditions. These may be found online. A few are listed in the following table:

Table 1. Coefficients for Use in Rohsenow's Correlation

fluid/surface	c
benzene/chromium	0.0101
butynol/copper	0.0030
carbon tetrachloride/copper	0.0130
ethanol/chromium	0.0027
isopropanol/copper	0.0025
pentane/chromium	0.0150
water/brass	0.0060
water/copper	0.0130
water/nickel	0.0060
water/stainless steel	0.0132

For the purposes of this illustration, we will consider water on stainless steel. The thermophysical properties from the previous example can be used, provided the water is at 36.875°C (98.38°F). The only additional property we need is the surface tension, σ, which at these conditions is about 700 dynes/cm or 0.7 N/m (0.004 lbf/inch).

---

[6] Rohsenow, W. M., "A Method of Correlation Heat Transfer Data for Surface Boiling of Liquids," Trans. ASME, Vol. 74, pp. 969-975, 1952.

Example 6. Calculations

	A	B	C	D	E	F
1			**Nucleate Boiling**			
2	Sample Calculations		SI Units		English Units	
3	INPUTS	symbol	units	value	units	value
4	sat. lqd. density	ρ_Y	kg/m³	995	lbm/ft³	62.12
5	sat. vap. density	ρ_g	kg/m³	0.0339	lbm/ft³	0.00211
6	gravity	g	m/s²	9.76	ft/sec²	32.018
7	viscosity	μ	cP	0.76	lbm/ft/hr	1.85
8	surface tension	σ	N/m	0.70	lbf/ft	0.048
9	latent heat	h_g	kJ/kg	2425	BTU/lbm	1043
10	specific heat	C	kJ/kg/°C	4.18	BTU/lbm/°F	0.998
11	thermal conduct.	k	W/m/°C	0.618	BTU/hr/ft/°F	0.357
12	temperature diff.	ΔT	°C	12	°F	21.6
13	Pr exponent	n	-	1	-	1
14	empirical coef.	c	-	0.0132	-	0.0132
15	SOLUTION					
16	Prandtl Number	Pr	-	5.17	-	5.17
17	heat flux	q	W/m²	6081	BTU/hr/ft²	1936
18	ht. tr. coef.	h	W/m²/°C	507	BTU/hr/ft²/°F	89.6
19						
20			user inputs in blue			
21			calculations in orange			

14

Example 7. The Boiling Curve

The concept of different boiling regimes was first introduced by Nukiyama[7], who observed this experimentally and proposed a curve similar to the one in this next figure:

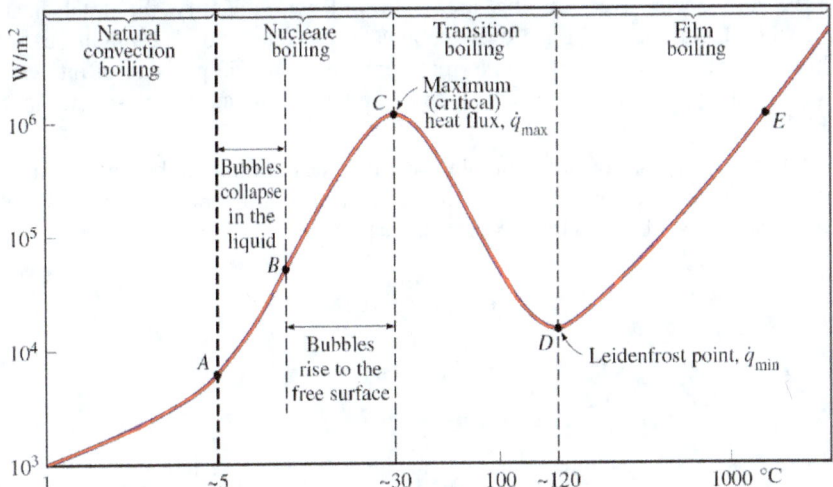

Figure 8. Boiling Curve for Water at Atmospheric Pressure

The horizontal axis is temperature difference (°C) and the vertical axis is heat flux (W/m²), both are logarithmic. Similar, though not identical, curves can be drawn for other liquids and pressures. The first region on the left (from approximately 1°C to 5°C) is called "nucleate convection boiling" and is characterized by the absence of bubbles. The next region (from about 5°C to 30°C) is called nuclear boiling and is characterized by the appearance of boiling, first forming, collapsing, and disappearing, then forming and rising to the surface as the temperature difference increases.

A maximum is seen at point C, while a minimum exists at point D. The maximum is called the "critical heat flux" because above this point the process responds in a somewhat counter-intuitive manner (i.e., greater temperature difference results in slower heat transfer rate). The Leidenfrost[8] point, named after the first experimentalist to report this phenomenon[9], is the onset of film

---

[7] Nukiyama, S., "The Maximum and Minimum Values of the Heat Transmitted from Metal to Boiling Water under Atmospheric Pressure," Journal Japan Society Mechanical Eng., Vol. 37, pp. 367, 1934 (Japanese); English Translation: Int. Journal of Heat Mass Transfer, Vol. 9, pp. 1419-1433, 1966.

[8] Johann Gottlob Leidenfrost (1715–1794) German physician and theologian.

[9] Leidenfrost, J. G., "On the Fixation of Water in Diverse Fire," 1756 (Latin); English Translation: Int. Journal of Heat and Mass Transfer, Vol. 9, pp. 1153-1166, 1966.

boiling in which there is a vapor film separating the liquid from the heating surface.

Some boiling processes, including those commonly occurring inside one or more tubes, experience a range of conditions, which will follow part of this curve and occur over varying areas. In order to quantify the total heat transferred, we must integrate over the area. To facilitate this, we might utilize this curve or one similar, created to represent the liquid and pressure of interest. We begin by digitizing this curve and putting it into an Excel spreadsheet (boiling_curve.xls).

For the purposes of this example, we will consider multiple boiling regimes along a single tube for which we know the surface temperature. We find the total heat transfer by integrating over the area:

$$\dot{Q} = \int \frac{d\dot{Q}}{dA} dA$$

(7.1)

For a single tube, the differential area is:

$$dA = \pi D \, dL$$

(7.2)

where D is the inside diameter and L is the length. When then have:

$$\dot{Q} = \pi D \int_0^{L:} \frac{d\dot{Q}}{dA} dL$$

(7.3)

As we have an empirical and not analytical expression for dQ/dA, analytical integration is not an option; therefore, we use numerical integration. Euler's method is adequate for such a task. The step size, ΔL, can be taken small enough to assure sufficient accuracy:

$$\dot{Q} = \pi D \sum_{i=1}^{n} \overline{\frac{d\dot{Q}}{dA}} \Delta L$$

(7.4)

16

The temperature differential along the length of the tube was curve-fitted from an experimental data set. The inside diameter (3 cm) came from the same data set, as did the tube length (1 m). The data and calculations may be found in the spreadsheet boiling_curve.xls and is shown in the figure below.

Example 7. Calculations

	A	B	C	D	E	F	G	H	I
1	boiling curve			tube calculations			inputs		
2	ΔT	q	L	ΔT	q	Q	D	cm	3
3	°C	W/m²	m	°C	W/m²	W	L	cm	100
4	1.00E+00	1.06E+03	0.00	2	1.77E+03	0			
5	1.26E+00	1.24E+03	0.05	5.4	5.84E+03	1.79E+03			
6	1.58E+00	1.47E+03	0.10	12	6.31E+04	1.80E+04			
7	2.00E+00	1.77E+03	0.15	22	5.08E+05	1.53E+05			
8	2.51E+00	2.18E+03	0.20	38	1.07E+06	5.24E+05			
9	3.16E+00	2.73E+03	0.25	59	3.30E+05	8.53E+05			
10	3.98E+00	3.52E+03	0.30	85	7.74E+04	9.49E+05			
11	5.01E+00	5.07E+03	0.35	118	2.76E+04	9.74E+05			
12	6.31E+00	8.49E+03	0.40	155	1.67E+04	9.84E+05			
13	7.94E+00	1.60E+04	0.45	196	1.70E+04	9.92E+05			
14	1.00E+01	3.34E+04	0.50	242	2.32E+04	1.00E+06			
15	1.26E+01	8.00E+04	0.55	290	3.22E+04	1.01E+06			
16	1.58E+01	1.74E+05	0.60	341	4.38E+04	1.03E+06			
17	2.00E+01	3.59E+05	0.65	393	5.83E+04	1.06E+06			
18	2.51E+01	6.93E+05	0.70	446	7.58E+04	1.09E+06			
19	3.16E+01	1.05E+06	0.75	499	9.61E+04	1.13E+06			
20	3.98E+01	1.01E+06	0.80	552	1.19E+05	1.18E+06			
21	5.01E+01	5.73E+05	0.85	604	1.45E+05	1.24E+06			
22	6.31E+01	2.54E+05	0.90	654	1.72E+05	1.32E+06			
23	7.94E+01	1.02E+05	0.95	703	2.01E+05	1.40E+06			
24	1.00E+02	4.43E+04	1.00	749	2.31E+05	1.51E+06			

The temperature difference, heat flux, and accumulated heat transfer along the tube are shown in the second figure.

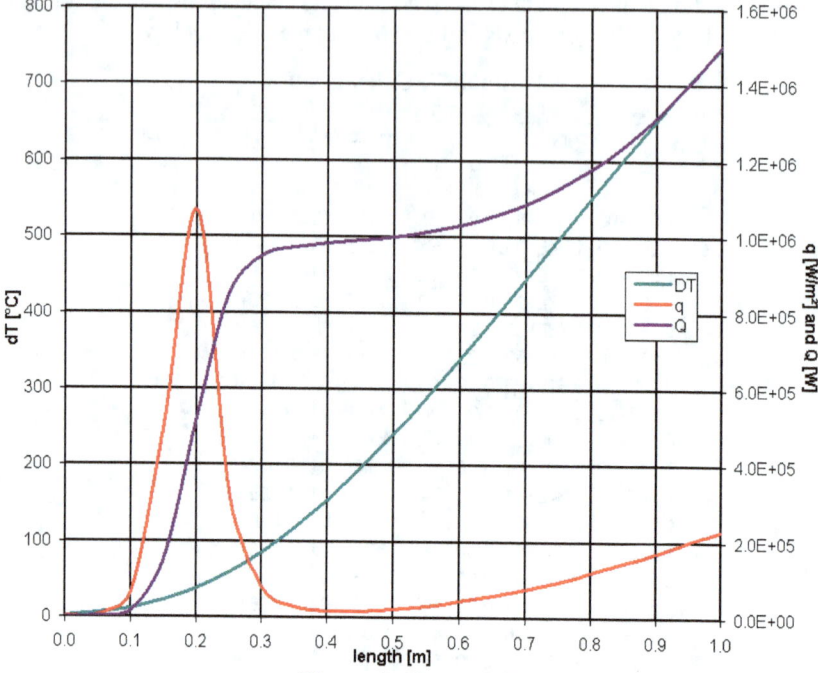

Figure 9. Example 7

Example 8. Pool Boiling

The foundational work on pool boiling is represented in the empirical Forster-Zuber[10] relationship:

$$\frac{q C_L \rho_L \sqrt{\pi \alpha_L}}{k_L h_{fg} \rho_V} \left(\frac{2\sigma}{\Delta p}\right)^{\frac{1}{2}} \left(\frac{\rho_L}{\Delta p}\right)^{\frac{1}{4}} = 0.0015 \, \mathrm{Re}_b^{0.62} \, \mathrm{Pr}_L^{\frac{1}{3}}$$

(8.1)

where q is the heat flux, C_L is the specific heat of the liquid, ρ_L is the density of the liquid, α_L is the thermal diffusivity of the liquid equal to the thermal conductivity divided by the density and specific heat ($\alpha = k/\rho C$), k_L is the thermal conductivity of the liquid, h_{fg} is the latent heat of vaporization, ρ_V is the density of the vapor, σ is the surface tension, Δp is the increase in (saturation) pressure due to the difference in temperature (ΔT) between the heating surface and boiling liquid, Re_b is the bubble Reynolds number, and Pr_L is the Prandtl number of the liquid. The bubble Reynolds number is given by:

$$\mathrm{Re}_b = \frac{\rho_L}{\mu_L} \left(\frac{C_L \rho_L \Delta T \sqrt{\pi \alpha_L}}{h_{fg} \rho_V}\right)^2$$

(8.2)

where μ_L is the dynamic viscosity of the liquid.

<u>Importance of Units</u>

It is essential when utilizing such formulae that the units are consistent. Dimensionless quantities, such a the Reynolds and Prandtl numbers, are useful here, as the units must all cancel out. We first consider Equation 8.2, where density is in mass/volume (m/L³), dynamic viscosity is in mass/length/time (m/Lt), specific heat is in energy/mass/temperature (E/mT), thermal diffusivity is in length²/time (L²/t), and latent heat is in energy/mass (E/m). Inserting these units into Equation 8.2 yields:

$$\frac{\left(\dfrac{m}{L^3}\right)}{\left(\dfrac{m}{Lt}\right)} \left[\frac{\left(\dfrac{E}{mT}\right)\left(\dfrac{m}{L^3}\right)(T)\sqrt{\left(\dfrac{L^2}{t}\right)}}{\left(\dfrac{E}{m}\right)\left(\dfrac{m}{L^3}\right)}\right]^2 = 1$$

(8.3)

We next consider the left side of Equation 8.1, where heat flux, q, has the units of power (energy per unit time) per unit area (E/L²t), surface tension, σ, has units of force per unit distance (F/L), thermal conductivity, k, has units of

---
[10] Forster, H. K. and N. Zuber, "Dynamics of Vapor Bubbles and Boiling Heat Transfer," AIChE Journal, Vol. 1, No. 4, p. 531, 1955.

power per unit length per temperature difference (E/tLT), and pressure has the units of force per unit area (F/L²). As this combination contains both energy (E) and force (F) we note that energy has the units of force times length (E=FL) and that force has the units of mass times length divided by time squared (F=mL/t²). Substituting these units into the left side of Equation 8.1 yields:

$$\frac{\left(\dfrac{E}{L^2 t}\right)\left(\dfrac{E}{mT}\right)\left(\dfrac{m}{L^3}\right)\sqrt{\left(\dfrac{L^2}{t}\right)}}{\left(\dfrac{E}{Lt}\right)\left(\dfrac{E}{m}\right)\left(\dfrac{m}{L^3}\right)}\left[\frac{\left(\dfrac{F}{L}\right)}{\left(\dfrac{F}{L^2}\right)}\right]^{\frac{1}{2}}\left[\frac{\left(\dfrac{m}{L^3}\right)}{\left(\dfrac{m}{Lt^2}\right)}\right]^{\frac{1}{4}} = 1$$

(8.4)

Do not presume when making calculations that SI units will all combine automatically without requiring powers of 10 to adjust grams, kilograms, poise, centipoise, pascals, kilopascals, joules, kilojoules, etc. I have seen huge and costly mistakes arise result from such presumption, for example: a pump that is 10x larger than required. It may be well worth the extra effort to perform the same calculations in both SI and English units as a check. If a design seems way too big or too small, it might well be a unit error. In this particular example, the only SI unit conversion required is to multiply kPa time 1000 to get Newtons/meter². For this calculation we can use many of the same properties as in Example 6. The results can be found in examples.xls.

Example 8. Calculations

	A	B	C	D	E	F
1			**Pool Boiling**			
2	Sample Calculations		SI Units		English Units	
3	INPUTS	symbol	units	value	units	value
4	sat. lqd. density	ρ_L	kg/m³	995	lbm/ft³	62.12
5	sat. vap. density	ρ_V	kg/m³	0.0339	lbm/ft³	0.00211
6	lqd. viscosity	μ_L	cP	0.76	lbm/ft/hr	1.85
7	surface tension	σ	N/m	0.70	lbf/ft	0.048
8	latent heat	h_{fg}	kJ/kg	2425	BTU/lbm	1043
9	lqd. specific heat	C_L	kJ/kg/°C	4.18	BTU/lbm/°F	0.998
10	lqd. thermal cond.	k_L	W/m/°C	0.618	BTU/hr/ft/°F	0.357
11	temperature diff.	ΔT	°C	20	°F	36
12	pressure diff.	ΔP	kPa	97.2	psia	14.1
13	CALCULATIONS					
14	thermal diffusivity	α	m²/s	1.49E-07	ft²/hr	0.00576
15	lqd. Prandtl	Pr_L	-	5.17	-	5.17
16	bubble Reynolds	Re_b	-	6.23E+05	-	6.24E+05
17	right side	-	-	10.2		10.2
18	left side/q	-	m²/W	6.76E-05	ft²hr/BTU	2.13E-04
19	heat flux	q	W/m²	1.50E+05	BTU/hr/ft²	47,678
20	ht. tr. coef.	h	W/m²/°C	7516	BTU/hr/ft²/°F	1324
21			user inputs in blue			
22			calculations in orange			

Example 9. Critical Heat Flux

The critical heat flux (upper inflection point in Figure 6) is an important consideration whenever boiling occurs, as this condition indicates a potential crisis: increasing surface temperature results in reduced heat transfer. This can result in burnout and in the case of reactor cooling, meltdown. Critical heat flux is an important consideration in design of heat exchangers. Stay well away from it. The empirical correlation of Zuber[11] was the first published for critical heat flux and subsequently modified by Lienhard and Dhir.[12]

$$q_{max} = 0.149 h_{fg} \rho_V \left[\frac{\sigma g (\rho_L - \rho_V)}{\rho_V^2} \right]^{\frac{1}{4}}$$

(9.1)

There are no new terms introduced in this equation and so we can evaluate it using the information used in Example 8. Properties for some other fluid can easily be substituted as needed.

Example 9. Calculations

	A	B	C	D	E	F
1	**Critical Heat Flux**					
2	Sample Calculations		SI Units		English Units	
3	INPUTS	symbol	units	value	units	value
4	sat. lqd. density	ρ_L	kg/m³	995	lbm/ft³	62.12
5	sat. vap. density	ρ_V	kg/m³	0.0339	lbm/ft³	0.00211
6	gravity	g	m/s²	9.76	ft/sec²	32.018
7	surface tension	σ	N/m	0.70	lbf/ft	0.048
8	latent heat	h_{fg}	kJ/kg	2425	BTU/lbm	1043
9	CALCULATIONS					
10	heat flux	q	W/m²	603,758	BTU/hr/ft²	191,390
11	user inputs in blue					
12	calculations in orange					

[11] Zuber, N., "Hydrodynamic Aspects of Boiling Heat Transfer," USAEC Report AECU-4439, 1959.
[12] Lienhard, J. H., and Dhir, V. K., "Extended Hydrodynamic Theory of the Peak and Minimum Heat Fluxes," NASA CR-2270, 1973.

Example 10. Minimum Heat Flux

Zuber also provided a slightly different formula for the minimum heat flux (lower inflection point in Figure 8):

$$q_{max} = 0.09 h_{fg} \rho_V \left[\frac{\sigma g (\rho_L - \rho_V)}{(\rho_L + \rho_V)^2} \right]^{\frac{1}{4}}$$

(10.1)

which we can also evaluate using the same properties in examples.xls:

Example 10. Calculations

	A	B	C	D	E	F
1	Minimum Heat Flux					
2	Sample Calculations		SI Units		English Units	
3	INPUTS	symbol	units	value	units	value
4	sat. lqd. density	ρ_L	kg/m³	995	lbm/ft³	62.12
5	sat. vap. density	ρ_V	kg/m³	0.0339	lbm/ft³	0.00211
6	gravity	g	m/s²	9.76	ft/sec²	32.018
7	surface tension	σ	N/m	0.70	lbf/ft	0.048
8	latent heat	h_{fg}	kJ/kg	2425	BTU/lbm	1043
9	CALCULATIONS					
10	heat flux	q	W/m²	2,128	BTU/hr/ft²	674
11	user inputs in blue					
12	calculations in orange					

Example 11. Double Pipe Heat Exchanger

The double pipe heat exchanger (shown schematically below) is perhaps the simplest to discuss but rarely used in practice. These don't provide much surface area and the rate of heat transfer is roughly proportional to surface area. They often appear in textbooks as an example.

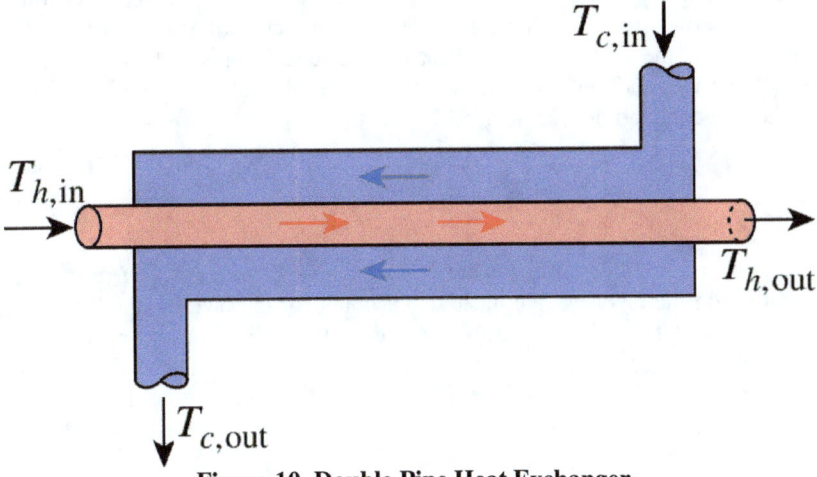

Figure 10. Double Pipe Heat Exchanger

While this figure shows the hot fluid flowing through the inside tube and the cold fluid flowing outside, this isn't necessarily the case. If the fluids are hotter than the surrounding environment and recovering the heat energy were of interest, this configuration makes more sense than the reverse. If the purpose were simply to remove heat, then passing the hotter fluid between the two pipes would also facilitate loss to the surroundings. If loss of heat to the surroundings is not desirable, the outer pipe can easily be insulated. Most often heat loss to the surroundings is ignored in example calculations, as is pressure drop, which there will clearly be for both fluids. We will discuss these details in subsequent examples.

This figure shows a counter-flow orientation of the two fluids. Unless there is phase change on one side or some unusual physical requirement, counter-flow is always preferred over parallel-flow. The mathematical reasoning for this is covered within any discussion of the LMTD (log-mean temperature difference) or the NTU-ε (number of transfer units - effectiveness) method. We first consider the simplified thermal circuit, represented as resistors in series:

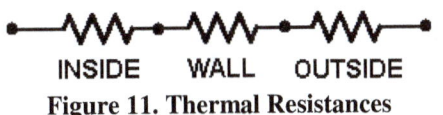

Figure 11. Thermal Resistances

We will solve this problem as a steady-state lumped system. This means we will calculate a single heat transfer coefficient for the inside and outside—in this case forced convection of a fluid without phase change (boiling or condensation)—and one for the tube wall. We will combine these as resistors in series and assume the resulting overall conductance, U, to be representative of the entire transfer surface. For this example we will use water for the cooling fluid (properties from the preceding examples) and ethanol as the hot fluid. We will calculate the inner and outer heat transfer coefficients as before from the Nusselt number and the tube wall resistance from the following equation, using the thermal conductivity for steel, k_W:

$$R_W = \frac{D_O}{2k_W} \ln\left(\frac{D_O}{D_I}\right)$$

(11.1)

As heat transfer coefficients are conductance (the reciprocal of resistance), we must add the inverses. We must also account for the fact that the inside and outside diameters of the inner tube represent two different areas (negligible in this particular case). This equation for conductance is formulated for use with the outside area of the inner pipe, so we adjust the inside heat transfer coefficient:

$$\frac{1}{U} = \frac{1}{h_o} + R_W + \frac{D_I}{D_O h_I}$$

(11.2)

The ethanol flows inside the inner pipe at a mean velocity of 1 m/s (3.28 ft/sec) and the water flows outside the inner pipe at a mean velocity of 1 m/s (3.28 ft/sec). The velocities should be high enough to assure turbulent flow (check the Reynolds numbers) and minimize surface buildup, while not so fast as to result in flow issues or erosion, which is rarely a problem with steel. We use the Dittus-Boelter correlation (Equation 4.1) for the ethanol but for the annulus (between the inner and outer pipe) we use the correlation of McAdams:[13]

$$Nu_{D_H} = 0.023 \, Re_{D_H}^{0.8} \, Pr^{\frac{1}{3}} \left(\frac{\mu}{\mu_w}\right)^{0.14}$$

(11.3)

where D_H is the hydraulic diameter of the annulus:

$$D_H = \frac{4A_C}{P_W} = D_O - D_I$$

(11.4)

---

[13] McAdams, W. H., *Heat Transmission 3rd Ed.*, McGraw-Hill, pp. 241-244, 1954.

where A_C is the cross-sectional area of the annulus, $A_C = \pi(D_O{}^2 - D_I{}^2)$, and P_W is the wetted perimeter, $P_W = \pi(D_O + D_I)$. The last terms is a correction used when the temperature differences are so large that the dynamic viscosity near the heated wall, μ_W, is significantly different than that in the core of the flow, μ. We will not use this correction in this example.

The inside diameter of the inner pipe is 2.66 cm (1.05 in). The outside diameter of the inner pipe is 3.34 cm (1.32 in). The inside diameter of the outer pipe is 7.79 cm (3.07 in). The ethanol must be cooled from 80°C (176°F) to 40°C (104°F). The water enters at 20°C (68°F). We see from the Example11 tab in examples.xls that the heat transfer is 43,289 W (147,709 BTU/hr) and that the exiting water temperature is 22.7°C (72.8°F). The required heat transfer area is 1.18 m² (12.7 ft²). This requires a length of 11.2 m (36.8 ft), which is why the double pipe design is rarely used. The inputs are listed below:

Example 11. Inputs

	A	B	C	D	E	F
			symbol	units	value	units	value
1				Double Pipe			
2	Sample Calculations			SI Units		English Units	
3	INPUTS		symbol	units	value	units	value
4	water inlet temp.		T_{c1}	°C	20	°F	68
5	water density		ρ_W	kg/m³	995	lbm/ft³	62.12
6	water sp. heat		C_W	kJ/kg/°C	4.18	BTU/lbm/°F	0.998
7	water th. cond.		k_W	W/m/°C	0.618	BTU/hr/ft/°F	0.357
8	water dyn. visc.		μ_W	cP	0.76	lbm/ft/hr	1.85
9	water velocity		V_W	m/s	1.00	ft/sec	3.28
10	ethanol inlet temp.		T_{H1}	°C	80	°F	176
11	ethanol outlet temp.		T_{H2}	°C	40	°F	104
12	ethanol density		ρ_O	kg/m³	789	lbm/ft³	49.26
13	ethanol sp. heat		C_O	kJ/kg/°C	2.46	BTU/lbm/°F	0.588
14	ethanol th. cond.		k_O	W/m/°C	0.170	BTU/hr/ft/°F	0.0983
15	ethanol dyn. visc.		μ_O	cP	0.500	lbm/ft/hr	1.21
16	ethanol velocity		V_W	m/s	1.00	ft/sec	3.28
17	inside pipe id		D_I	cm	2.66	in	1.05
18	inside pipe od		D_O	cm	3.34	in	1.32
19	outside pipe id		D_D	cm	7.79	in	3.07
20	steel therm. cond.		k_S	W/m/°C	45.0	BTU/hr/ft/°F	26.02

The heat transfer, area, conductance, and temperature difference are related by:

$$\dot{Q} = UA \cdot LMTD \qquad (11.5)$$

The calculations are listed below:

Example 11. Calculations

	A	B	C	D	E	F
21	CALCULATIONS					
22	Re inside pipe	Re_I	-	42,045	-	42,045
23	ethanol Prandtl	Pr_I	-	7.24	-	7.23
24	ethanol Nusselt	Nu_I	-	208		208
25	ethanol ht. tr. co.	h_I	W/m²/°C	1,329	BTU/hr/ft²/°F	234
26	ethanol mass flow	m_E	kg/s	0.440	lbm/hr	3,492
27	ht. tr. from ethanol	Q_E	W	43,289	BTU/hr	147,709
28	hydraulic diameter	D_H	cm	4.45	in	1.75
29	Re outside pipe	Re_o	-	57,948	-	57,948
30	water Prandtl	Pr_o	-	5.17	-	5.17
31	water Nusselt	Nu_o	-	257	-	257
32	water ht. tr. co.	h_o	W/m²/°C	3,568	BTU/hr/ft²/°F	629
33	pipe wall ht. tr. co.	h_W	W/m²/°C	11,923	BTU/hr/ft²/°F	2,101
34	overall conduct.	U	W/m²/°C	1,037	BTU/hr/ft²/°F	183
35	water mass flow	m_W	kg/s	3.87		30,744
36	water exit temp.	T_{cz}	°C	22.7	°F	72.8
37	water temp. rise	ΔT	°C	2.67	°F	4.81
38	LMTD	ΔT	°C	35.4	°F	63.8
39	required cond.	UA	W/°C	1,221	BTU/hr/°F	2,315
40	required area	A	m²	1.18	ft²	12.7
41	required length	L	m	11.2	ft	36.8

There is a check for Reynolds number >4000 (turbulent) in the spreadsheet. When making calculations for other fluids, temperatures, or geometries it is essential to make sure the flow inside and outside the pipe is turbulent, as the correlations do not apply to the laminar or transitional flow regimes.

Example 12. Finned Tube Heat Exchanger

Finned tube heat exchangers are far more common than double pipe. The fins are most often slid onto a tube and possibly crimped (types 1 and 2 below). Machined or cast fins (type 3 below) are much more expensive and only used in special applications where appearance is more important than cost-effectiveness. These fins are external. We will consider internal ones in the next chapter. Fins are typically added on the side where the convective heat transfer coefficient is lowest. Equation 11.2 (adding thermal resistances in series) reveals that this is a *weakest link* sort of problem. Improvements to a high-conductance (low-resistance) element yield little or no impact; whereas, improvements to a low-conductance (high-resistance) element can yield a significant increase in heat transfer. This is why we often see such fins exposed to air on the outside of tubes through which flow a liquid. The thermal conductivity of liquids are often much higher than the thermal conductivity of air.

Figure 12. Types of Tube Fins

The impact of external fins is most often expressed as an effective surface area or an efficiency applied to the total surface area. As the fins are most often spaced fairly close together, the end result will be somewhat less than were the same area equally exposed to the surroundings receiving the heat; thus the fin efficiency is ≤ 1.

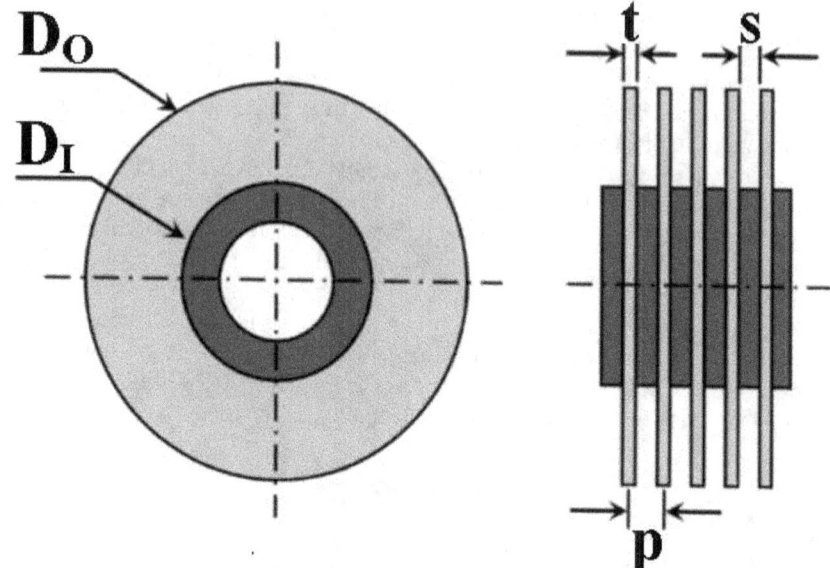

Figure 13. Fin Dimensions

Here D_I is the inside (or root) diameter of the fin, D_O is the outside diameter, t is the thickness, s is the spacing, and p is the pitch. The fins should be in good contact with the tube wall, but this is not always possible or practical, especially when corrosion is present. Thermal resistance between the fins and tube wall may be accounted for using an empirical factor, which should consider the fabrication, materials, and aging. While not all fins have a uniform cross-section (some are tapered), we will only consider those illustrated above.

Our first example in this chapter will take the traditional approach. The next example will follow more recent developments. We begin with the Biot[14] number, which is the dimensionless ratio of convective to conductive heat transfer coefficient (or the ratio of conductive to convective thermal resistance).

$$Bi = \frac{hl}{k}$$

(12.1)

where h is the convective heat transfer coefficient, l is the characteristic length of the fin (equal to the area divided by the perimeter or $t/2$ for rectangular cross-section), and k is the thermal conductivity (of the fin). For this example we will consider liquid R134a (1,1,1,2-tetrafluoroethane, a common refrigerant) flowing inside a copper tube of 10.2 mm (0.402 in) diameter, having a wall thickness of 1.24 mm (0.049 in) and an outside diameter of 12.7 mm (0.5 in). Aluminum fins

---

[14] Jean-Baptiste Biot (1774–1862) French mathematician and physicist who studied magnetism and light.

of 31.8 mm (1.25 in) outer diameter, 0.794 mm (0.0314 in) thickness, and 0.476 mm (0.188 in) spacing, resulting in a 5.55 mm (0.219 in) pitch. The thermal conductivity of copper is 398 W/m/°C (230 BTU/hr/ft/°F). A typical heat transfer coefficient for natural convection in air across a horizontal tube is 10 W/m²/°C (1.76 BTU/hr/ft²/°F). The Biot number is then 9.97×10^{-6} (dimensionless), which is often quite small in such cases. The fin efficiency in this case is given by the following:

$$\eta_F = \frac{\tanh(\varepsilon)}{\varepsilon}$$

$$\varepsilon = \frac{(L+l)\sqrt{Bi}}{l} \qquad (12.2)$$

where the quantity L is equal to the length of the fins. The characteristic length of the fins in this case is 0.397 mm (0.0156 in) and the physical length is 19.1 mm (0.75 in), making the parameter ε equal to 0.155 (dimensionless) and the fin efficiency 0.992 (dimensionless).

We calculate the heat transfer coefficient (reciprocal of resistance) for the tube wall as before using Equation 11.1 to get 2.87×10^5 W/m²/°C (5.06×10^4 BTU/hr/ft²/°F).

The density of R134a liquid is about 700 kg/m³ (43.7 lbm/ft³). The specific heat of the liquid is 1.96 kJ/kg/°C (0.468 BTU/lbm/°F). The thermal conductivity of the liquid is 0.0655 W/m/°C (0.0379 BTU/hr/ft/°F). The dynamic viscosity of the liquid is 0.135 cP (0.327 lbm/ft/hr), making the Prandtl number equal to 4.04 (dimensionless). If the mean velocity inside the tube is 1 m/s (3.28 ft/sec), then the Reynolds number will be 52,945 (fully turbulent). We use the Dittus-Boelter correlation (Equation 4.1) to calculate the Nusselt number for forced convection inside the tube and obtain a value of 220 (dimensionless). The heat transfer coefficient for turbulent forced convection inside the tube is then 1,136 W/m²/°C (200 BTU/hr/ft²/°F).

Finally, we use the reciprocal relationship (Equation 11.2) to calculate the overall heat transfer coefficient, U, 9.85 W/m²/°C (1.73 BTU/hr/ft²/°F). Multiplying this by the surface area of the fins and the log-mean temperature difference will yield the rate of heat transfer. Given design flow rates and temperatures, the required tube length can be calculated. These same calculations in the spreadsheet (tab Example12) can be used with different dimensions and properties to solve similar problems. Always check the Reynolds number to be sure the flow inside the tube is turbulent.

Example 12. Calculations

	A	B	C	D	E	F
1			**Externally Finned Tube**			
2	Sample Calculations			SI Units	English Units	
3	INPUTS	symbol	units	value	units	value
4	tube outside dia.	D_o	mm	12.7	in	0.5000
5	tube wall thick.	t_w	mm	1.24	in	0.0490
6	tube inside dia.	D_I	mm	10.2	in	0.4020
7	fin diameter	D_f	mm	31.8	in	1.2500
8	fin thickness	f_f	mm	0.794	in	0.0313
9	fin spacing	s	mm	4.76	in	0.1875
10	fin pitch	p	mm	5.56	in	0.2188
11	copperl therm. cond.	k_s	W/m/°C	398	BTU/hr/ft/°F	230
12	outside conv. ht. tr.	h_c	W/m²/°C	10.0	BTU/hr/ft²/°F	1.76
13	R134a density	ρ	kg/m³	700	lbm/ft³	43.7
14	R134a sp. heat	C	kJ/kg/°C	1.96	BTU/lbm/°F	0.468
15	R134a th. cond.	k	W/m/°C	0.0655	BTU/hr/ft/°F	0.0379
16	R134a dyn. visc.	μ	cP	0.135	lbm/ft/hr	0.327
17	R134a velocity	V	m/s	1.00	ft/sec	3.28
18	CALCULATIONS					
19	Biot	Bi	-	9.97E-06	-	9.97E-06
20	fin length	L	mm	19.1	in	0.750
21	characteristic length	l	mm	0.397	in	0.0156
22	fin factor	ε	-	0.155	-	0.155
23	fin efficiency	η	-	0.992	-	0.992
24	tube wall ht. tr. co.	h_w	W/m²/°C	2.87E+05	BTU/hr/ft²/°F	5.06E+04
25	Reynolds number	Re	-	52,945	-	52,945
26	Prandtl number	Pr	-	4.04	-	4.04
27	Nusselt number	Nu	-	220	-	220
28	heat transfer coef.	h_I	W/m²/°C	1,136	BTU/hr/ft²/°F	200
29	overall conduct.	U	W/m²/°C	9.85	BTU/hr/ft²/°F	1.73
30			user inputs in blue			
31			calculations in orange			

Example 13. Enhanced Surfaces

Enhanced surfaces, including internal fins, can be quite effective. Wolverine makes a wide assortment of enhanced tubes and also provides a useful handbook on how to make the necessary design calculations.[15] This resource is available free online.[16] In the Wolverine handbook you will often see references to the friction factor, f, given by the Darcy-Weisbach equation:[17,18]

$$\Delta p = f\left(\frac{L}{d}\right)\left(\frac{\rho V^2}{2g}\right)$$

(13.1)

where Δp is the pressure drop (for example, in a pipe of diameter, d, and length, L), filled with a liquid of density, ρ, flowing at a mean velocity of, V. The acceleration of gravity is indicated by, g. Note that this is the *Darcy* friction factor and not the Fanning. The two differ by a factor of 4. The friction factor is often associated with the Moody chart:[19]

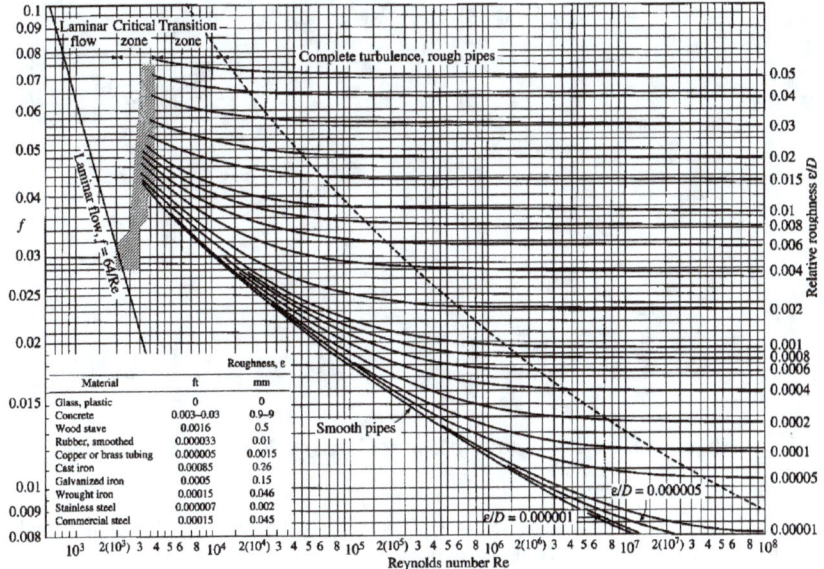

Figure 14. Moody Chart

[15] Wolverine Tube, Inc. website https://wlv.com/
[16] Thome, J. R., *Engineering Data Book II*, Wolverine, 2009.
[17] Henry Philibert Gaspard Darcy (1803–1858) French engineer who made important advances in the study of hydraulics and flow in porous media.
[18] Julius Ludwig Weisbach (1806-1871) German mathematician and engineer.
[19] Lewis Ferry Moody (1880–1953) American engineer and inventor; the first Professor of Hydraulics in the School of Engineering at Princeton.

The Colebrook-White[20] empirical relationship for friction factor in pipes of varying roughness, ε/d, is given by the following equation and agreement shown in the following figure:

$$\frac{1}{\sqrt{f}} = -2\log_{10}\left(\frac{\varepsilon}{3.7d} + \frac{2.51}{\mathrm{Re}\sqrt{f}}\right)$$

(13.2)

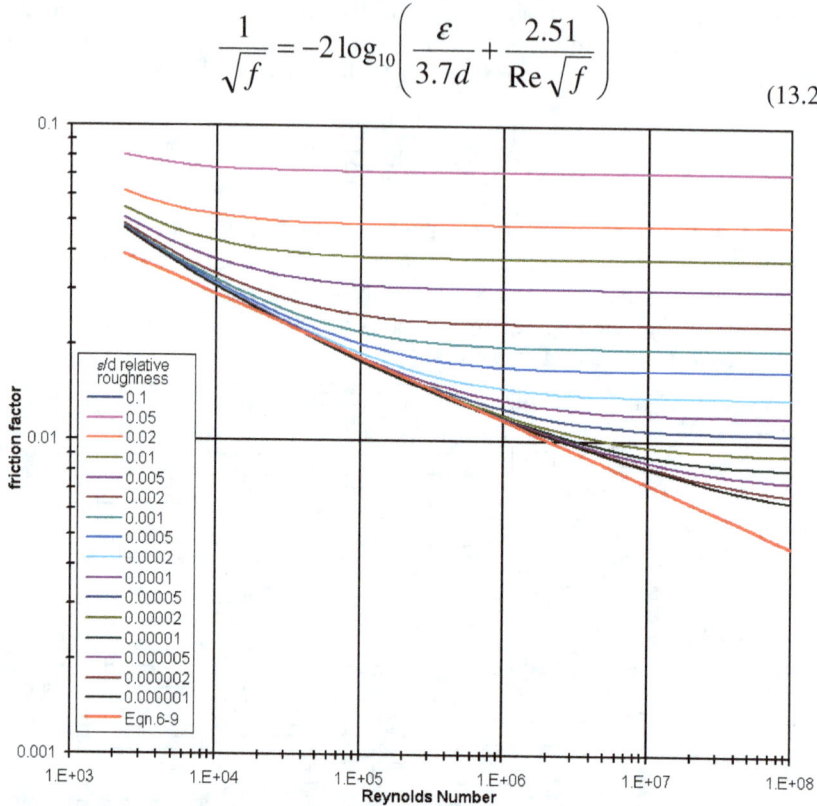

Figure 15. Friction Factor Using Colebrook-White Formula

The friction factor is used to calculate pressure drop (an important consideration in heat exchanger design) and also comes into the heat transfer calculations by way of Reynolds Analogy:[21]

$$St = \frac{Nu}{\mathrm{Re}\,\mathrm{Pr}} = \frac{f}{2}$$

(13.3)

[20] Colebrook, C. F. and White, C. M., "Experiments with Fluid Friction in Roughened Pipes". Proceedings of the Royal Society of London, Series A, Mathematical and Physical Sciences, Vol. 161, No. 906, pp. 367–381, 1937.
[21] Reynolds, O., *Scientific Papers of Osborne Reynolds*, Vol. II, Cambridge University Press, 1901.

where *St* is the Stanton number. Solving for the heat transfer coefficient, we get:

$$h = f\left(\frac{\rho CV}{2}\right) = \Delta p\left(\frac{dgC}{LV}\right)$$

$$h \propto f \propto \Delta p \tag{13.4}$$

Thus the heat transfer coefficient, h, is roughly proportional to the friction factor, f, which is proportional to the pressure drop, making $h \propto f \propto \Delta p$. This means that, if we do something to increase the pressure drop (add roughness, bumps, fins, grooves, or twisted tapes), we can expect a proportional increase in heat transfer coefficient.

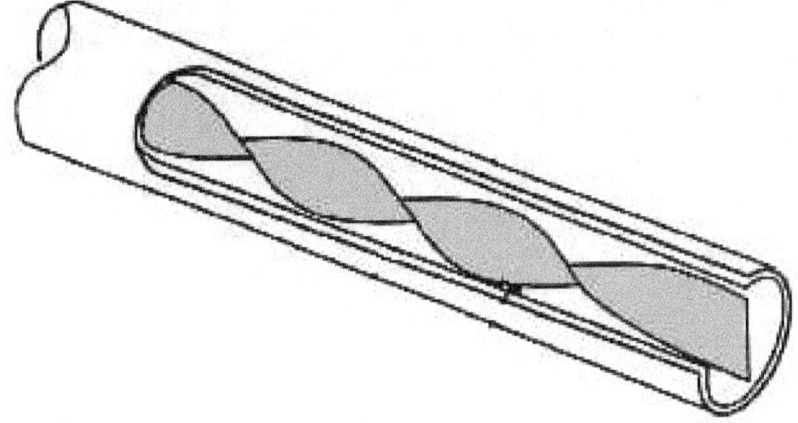

Figure 16. Twisted Tape Tube Insert

This, of course, only applies to the process impacted; that is, the internal forced convection heat transfer coefficient when the modifications are inside the tube. Such enhancements do not change the thermal resistance of the tube wall, nor the heat transfer coefficient on the outside of the tubes. Therefore, the only designs where it is cost-effective to employ enhanced surfaces such as twisted tapes are when the convective heat transfer is relatively low. This would occur if the velocity of the fluid inside the tube were low (possibly approaching laminar but not fully turbulent) or when the thermal conductivity of the fluid is low (which is the case with some hydrocarbons and other common industrial fluids).

Using Equation 13.3 we can calculate the impact on friction factor and turbulent convective heat transfer coefficient for different tube roughness (see Example13 tab in examples.xls):

Example 13. Calculations

	A	B	C	D	E
1	**Enhanced Surface Roughness**				
2	Sample Calculations		relative roughness, ε/d		
3	Variable	symbol	0.0002	0.002	0.02
4	Reynolds number	Re	5.00E+05	5.00E+05	5.00E+05
5	friction factor	f	0.015	0.024	0.049
6	presure drop	Δp/Δp	1.000	1.541	3.154
7	heat transfer coef.	h/h	1.000	1.541	3.154
8	**user inputs in blue**				
9	**calculations in orange**				

Here we see that increasing the relative roughness, ε/d, by a factor of 10x increases the pressure drop and turbulent convective heat transfer coefficient by a factor of 1.541 and by 100x increases the same parameters by a factor of 3.154. This is why simple roughness may not be adequate, leading to more flow-disruptive measures such as bumps, grooves, and twisted tapes.

Example 14. Water-Cooled Steam Surface Condenser

The two key references for water-cooled steam surface condensers are: Heat Exchange Institute Standards for Steam Surface Condensers 11th Ed. (2012) and ASME PTC-12.2 (2010). These two documents are quite different in approach and content, the former is more academic and the latter is more practical. PTC-12.2 presents dimensionless correlations involving Reynolds and Nusselt numbers and the HEI document contains many empirical curves. The primary concern of these documents and various formulas therein is to obtain a value for the overall heat transfer coefficient, U. Both contain guidelines for the physical design, including structural and practical considerations, such as air removal systems. Here we will only consider the thermal aspects of these two documents.

HEI Method

We begin with the HEI method, which is entirely in English units. The document does include a complete list of conversion factors in Appendix B. As the English tables and curves may be found in the HEI document, which is available free online, the SI equivalents will be provided herein. The uncorrected overall heat transfer coefficient, U, as a function of tube diameter (in mm) and water velocity (in m/s) are shown in this figure:

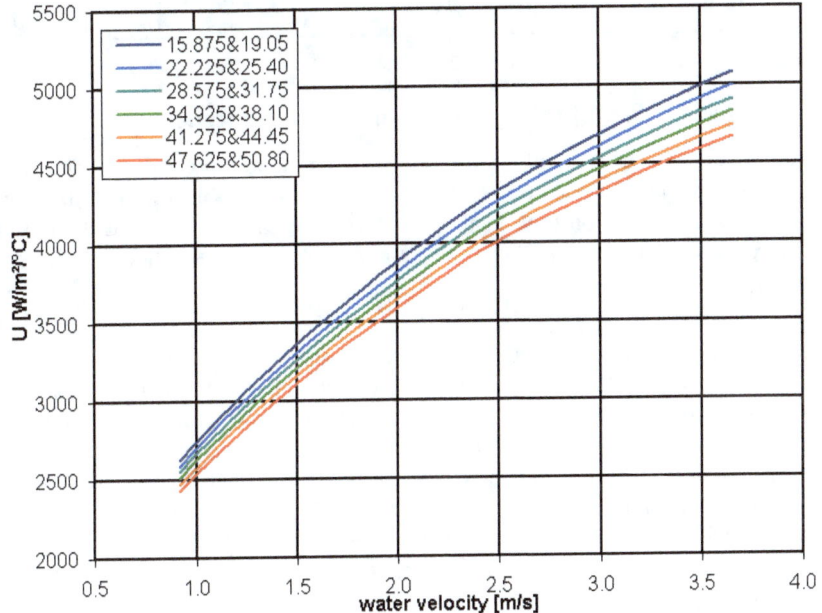

Figure 17. HEI Uncorrected Overall Heat Transfer Coefficient

You will find a bivariate curve-fit of this function in spreadsheet HEI.xls to facilitate your analyses.

The HEI procedure calls for several multiplicative corrections to be applied, the first of these being the inlet water temperature correction factor, which is unitless.

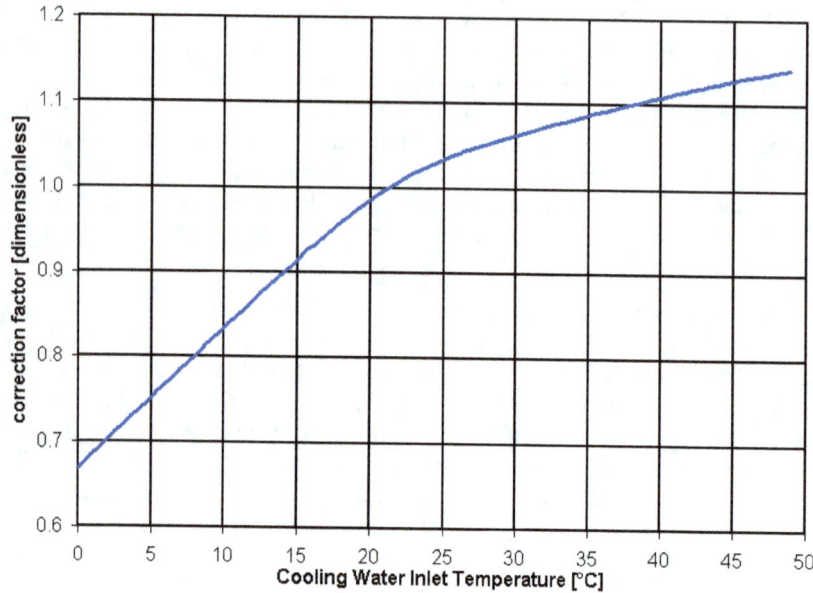

Figure 18. HEI Inlet Water Temperature Correction Factor

There is also a curve-fit of this function in the same spreadsheet. The next correction in the HEI method is the material factor. This is simply a table of unitless values based on tube material and wall thickness (shown at the top of the next page). The corrected overall heat transfer coefficient is then calculated:

$$U_{COR} = U_{UNC} F_W F_M C_F \tag{14.1}$$

where U_{COR} is the corrected overall heat transfer coefficient, U_{UNC} is the uncorrected, F_W is the water temperature correction factor, F_M is the material and gauge correction factor, and C_F is the cleanliness factor. The cleanliness factor should be between 75% and 100% with a design value of about 90%, which is equivalent to a 10% performance margin. If the cleanliness is below 75% (which is not uncommon in operating plants), the condenser should be cleaned. Different types of cleaning with wide ranging level of effort may be required, depending on the operating conditions, age, and water cleanliness.

Table 2. HEI Tube Material and Gauge Correction Factor (unitless)

	Birmingham Wire Gauge (BWG)								
	25	24	23	22	20	18	16	14	12
	Tube Wall Thickness [mm]								
material	0.508	0.559	0.635	0.711	0.888	1.245	1.651	2.108	2.769
Admiralty Metal	1.03	1.03	1.02	1.02	1.01	1.00	0.98	0.96	0.93
Arsenical Copper	1.04	1.04	1.04	1.03	1.03	1.02	1.01	1.00	0.98
Copper Iron 194	1.04	1.04	1.04	1.04	1.03	1.03	1.02	1.01	1.00
Aluminum Brass	1.03	1.02	1.02	1.02	1.01	0.99	0.97	0.95	0.92
Aluminum Bronze	1.02	1.02	1.01	1.01	1.00	0.98	0.96	0.93	0.89
90-10 Cu-Ni	1.00	0.99	0.99	0.98	0.96	0.93	0.89	0.85	0.80
70-30 Cu-Ni	0.97	0.97	0.96	0.95	0.92	0.88	0.83	0.78	0.71
Carbon Steel	1.00	1.00	0.99	0.98	0.97	0.93	0.89	0.85	0.80
SS 304/316	0.91	0.90	0.88	0.86	0.82	0.75	0.69	0.62	0.54
Titanium	0.95	0.94	0.92	0.91	0.88	0.82	0.77	0.71	0.63
UNS N08367	0.90	0.89	0.87	0.85	0.81	0.74	0.67	0.60	0.52
UNS S43035	0.95	0.94	0.92	0.91	0.88	0.82	0.77	0.71	0.63
UNS S44735	0.93	0.91	0.90	0.88	0.85	0.78	0.72	0.65	0.57
UNS S44660	0.93	0.91	0.90	0.88	0.85	0.78	0.72	0.65	0.57

The tube length for large water-cooled surface condensers is often determined by the size of the low-pressure steam turbine, which may sit directly above and discharge down across the tubes. Other factors controlling tube length include availability and the size of railroad boxcars used for transport. It is for this last reason that 40 foot (12.2 m) tubes are often used. There will be thousands of such tubes in a large condenser so that common sizes are often selected, for example 1 inch (25.4 mm) inside diameter. The tube material is selected next, which may be driven by water quality. In this example, we will assume ocean water will be used for cooling so that titanium tubes are preferable. With the diameter and material selected, the wall thickness follows (see HEI document) 22 BWG (0.711 mm).

The target water velocity is also dependent on the tube size and material (as described in the HEI document) and will be 1.8 m/s (5.91 ft/sec) for our example calculations. The design inlet water temperature is 26°C (78.8°F). From Figure 17 we get an uncorrected overall heat transfer coefficient of 3599 W/m²/°C (633.9 BTU/hr/ft²/°F). From Figure 18 we get a water temperature correction factor of 1.04 (unitless). From Table 2 we get a material and gauge factor of 0.91 (unitless). With a design cleanliness of 90% this yields a corrected overall heat transfer coefficient of 3,069 W/m²/°C (540.5 BTU/hr/ft²/°F).

The log-mean temperature difference is also selected during the design process. A typical value is 10°C (18°F). The steam mass flow rate is specified by the purchaser and will be assumed in this case to be 907,185 kg/hr (2,000,000 lbm/hr). The latent heat of the condensing steam is also specified by the purchaser (or steam turbine manufacturer) and will be set to 2,256 kJ/kg (970

BTU/lbm) for this example. The total heat transfer from the steam (i.e., condenser duty) is then 568,558 kW (1.94×10^9 BTU/hr).

The required surface area is equal to the duty divided by the overall heat transfer coefficient and log-mean temperature difference, which calculation yields 18,526 m² (199,408 ft²). The tube wall thickness is 0.711 mm (0.028 in), making the outer diameter of the tubes 26.8 mm (1.06 in). With a tube length of 12.2 m (40 feet), the required number of tubes would be 18,033.

Example 14. HEI Method Calculations

	A	B	C	D	E	F
1	**Steam Surface Condenser Using HEI Method**					
2	Sample Calculations			SI Units	English Units	
3	INPUTS	symbol	units	value	units	value
4	tube inside dia.	D_I	mm	25.4	in	1.00
5	tube thickness	t	mm	0.711	in	0.0280
6	tube length	L	m	12.2	ft	40
7	water velocity	V	m/s	1.8	ft/sec	5.91
8	water inlet temp.	Tin	°C	26	°F	78.8
9	LMTD	LMTD	°C	10	°F	18
10	cleanliness	C_f	-	90%	-	90%
11	steam flow	m_s	kg/hr	907,185	lbm/hr	2,000,000
12	latent heat	hfg	kJ/kg	2,256	BTU/lbm	970
13	CALCULATIONS					
14	uncorrected ht. tr.	U	W/m²/°C	3,599	BTU/hr/ft²/°F	633.9
15	water temp. corr.	F_W	-	1.04	-	1.04
16	material factor	F_M	-	0.91	-	0.91
17	corrected ht. tr.	U	W/m²/°C	3,069	BTU/hr/ft²/°F	540.5
18	heat transfer (duty)	Q	kW	568,558	BTU/hr	1.94E+09
19	required area	A	m²	18,526	ft²	199,408
20	tube outside dia.	D_o	mm	26.82	in	1.06
21	number of tubes	N	-	18,033	-	18,033
22	user inputs in blue					
23	calculations in orange					

ASME PTC-12.2 Method

We will use the same heat load and geometry (tube material, diameter, thickness, and length) as before, only applying the calculations contained in the ASME document. The tube wall thermal resistance is given by Equation 11.1 and the thermal conductivity of titanium is 11.4 W/m/°C (6.59 BTU/hr/ft/°F), making the heat transfer coefficient (reciprocal of thermal resistance) for the tube wall 15,605 W/m²/°C (2,750 BTU/hr/ft²/°F).

For turbulent convective heat transfer inside tubes, the ASME method uses the correlation of Rabas and Cane[22], which is only trivially different from that of

---

[22] Rabas, T. J., and Cane, D., "An Update of Intube Forced Convection Heat Transfer Coefficients of Water," *Desalinization*, Volume 44, pp. 109–119, 1983.

Dittus-Boelter, one of many examples of why the ASME method might be called, *Much Ado About Nothing*.

$$Nu = 0.0158 Re^{0.835} Pr^{0.462} \tag{14.2}$$

Using the inlet temperature, the density of water is 997 kg/m³ (62.2 lbm/ft³) and the dynamic viscosity is 1 cP (2.42 lbm/ft/hr). Combining these and the mean velocity, we obtain a Reynolds number inside the tubes of 45,583 (fully turbulent). The specific heat is 4.18 kJ/kg/°C (1 BTU/lbm/°F). The Prandtl number is then 6.99. Using Equation 14.2 we obtain a Nusselt number of 301 and a turbulent convective heat transfer coefficient of 7,094 W/m²/°C (1,250 BTU/hr/ft²/°F).

One of the more humorous parts of ASME PTC-12.2 (2010) is Section 5-2.4 Shellside Resistance (quoted below):

> The shellside condensing heat transfer is the most complex component in the evaluation of a steam surface condenser. Numerous correlations of the Nusselt equation have been developed for the film coefficient for various condensing situations. These correlations are based on specific condensing conditions requiring a detailed knowledge of the shell and tube nest geometries and the condensing conditions, and resulting in significantly different prediction values of the heat-transfer coefficient. As a result, the test value of the shellside resistance is determined in **para. 5-1.6** based on the difference between the test-calculated overall heat-transfer coefficient and sum of the calculated values for the other heat transfer resistances.

the "funny" part being that there is no Section (paragraph) 5-1.6 in that document—or in the previous 1998 document either. The calculation alluded to in the paragraph above (i.e., Nusselt's) is Equation 5.1 so that we can use the heat transfer coefficient from Example 5 of 9074 W/m²/°C (1599 BTU/hr/ft²/°F).

Combining these factors using Equation 11.2 yields an overall conductance of 3,094 W/m²/°C (545 BTU/hr/ft²/°F), a required surface area of 18,373 m² (197,676 ft²), and a required number of tubes of 17,890. With all the complicated description and calculations, the end result differs by less than 1%.

Example 14. ASME Method Calculations

	A	B	C	D	E	F
1	**Steam Surface Condenser Using ASME Method**					
2	Sample Calculations		SI Units		English Units	
3	INPUTS	symbol	units	value	units	value
4	tube inside dia.	D_I	mm	25.4	in	1.00
5	tube thickness	t_o	mm	0.711	in	0.0280
6	tube conductivity	k_T	W/m/°C	11.4	BTU/hr/ft/°F	6.59
7	tube length	L	m	12.2	ft	40
8	water velocity	V	m/s	1.80	ft/sec	5.91
9	water density	ρ_w	kg/m³	997	lbm/ft³	62.2
10	water specific heat	C_w	kJ/kg/°C	4.18	BTU/lbm/°F	1.00
11	water dyn. visco.	μ_w	cP	1.00	lbm/ft/hr	2.42
12	water thermal cond.	k_w	W/m/°C	0.598	BTU/hr/ft/°F	0.346
13	water inlet temp.	Tin	°C	26	°F	78.8
14	LMTD	LMTD	°C	10	°F	18
15	steam flow	m_s	kg/hr	907,185	lbm/hr	2,000,000
16	latent heat	hfg	kJ/kg	2,256	BTU/lbm	970
17	CALCULATIONS					
18	tube outside dia.	D_o	mm	26.8	in	1.06
19	tube wall ht. tr. co.	h_w	W/m²/°C	15,605	BTU/hr/ft²/°F	2,750
20	Reynolds Number	Re	-	45,583	-	45,583
21	Prandtl Number	Pr	-	6.99	-	6.99
22	Nusselt Number	Nu	-	301	-	301
23	tube inside ht. tr.	h_T	W/m²/°C	7,094	BTU/hr/ft²/°F	1,250
24	condensation ht. tr.	h	W/m²/°C	9074	BTU/hr/ft²/°F	1599
25	overall ht. tr. co.	U	W/m²/°C	3,094	BTU/hr/ft²/°F	545
26	heat transfer (duty)	Q	kW	568,558	BTU/hr	1.94E+09
27	required area	A	m²	18,373	ft²	197,676
28	number of tubes	N	-	17,884	-	17,876
29	user inputs in blue					
30	calculations in orange					

Example 15. Air-Cooled Steam Surface Condenser

Large air-cooled steam surface condensers (ACCs) are quite different from water-cooled ones (WCCs) in appearance, performance, and documentation. For example, WCC manufacturers always list the total surface area, number, and size of tubes; whereas, ACC manufacturers rarely provide such information. WCC designs never list the power required to pump water through them; whereas, ACC designs always list the fan power required. WCC designs are typically a large rectangular steel box; whereas, a variety of designs are used for ACCs, for example:

Figure 19. Typical Air-Cooled Steam Condenser

The fan(s) may be above or below the heat exchangers, drawing air in or pushing air out through the banks of finned tubes. Discussions abound on which is preferable for various reasons, mostly related to accumulation of debris, maintenance, and bypass (air traveling through gaps and avoiding the heat transfer surfaces). The performance is similar and design varies from one manufacturer to another.

Burping an ACC

An essential functional aspect of these devices is that the steam must condense and flow to the collection area without accumulating anywhere along the path. Also, non-condensables (mostly air) must naturally flow to and collect at some convenient location so as to facilitate removal. Failure to adequately remove non-condensables can have a huge detrimental impact on performance. Removal of air from an ACC is called *burping*. I was performing an acceptance on one such device—which obviously needed burping—but the factory representative on site had never heard of this and had no idea what I was talking about. At first, the condenser was failing badly. After burping, it performed more than 2% above guarantee. This was an unusual (and laughable) situation, as most manufacturers are knowledgeable about their products.

The flow of steam and air are shown schematically below:

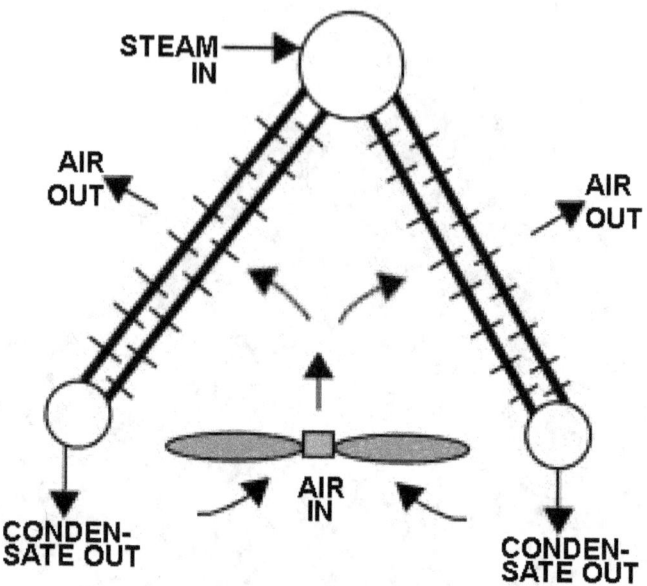

Figure 20. ACC Schematic

As we saw in Example 12, the thermal resistance is dominated by the turbulent convective transport with the air. Condensation of steam inside tubes has a somewhat lower heat transfer coefficient than outside tubes but this is still orders of magnitude greater than the turbulent forced convection. Thermal conductance of the tube wall and fins is orders of magnitude greater still. We therefore focus on the convective contribution.

Tube and fin materials vary from one design to another in consideration of several factors, such as whether or not the unit is near the sea, as salt air is very corrosive. Fin design often considers accumulation of debris and ease of cleaning. Many fins (such as a car radiator or window air conditioner) could not survive pressure washing. Some fin designs are proprietary and protected by patents. Perhaps this is why some manufacturers provide very little physical information about an ACC.

The minute details of forced air turbulent convection have been studied and presented in technical papers but this information is rarely used by industry. Instead, empirical correlations (based on laboratory experiments) are most often used. Several of these may be found online. One that often appears at the top of a search is on http://www.engineeringtoolbox.com/:

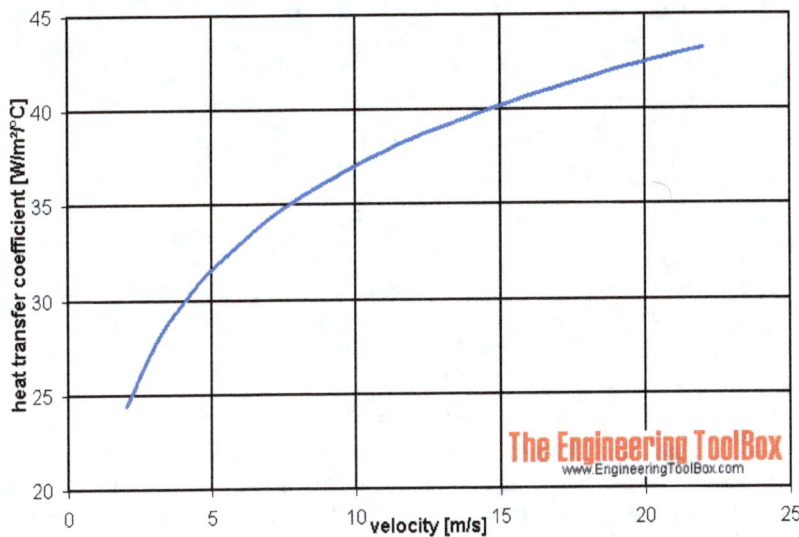

Figure 21. Forced Air Convection Heat Transfer Coefficient

A curve-fit and macro may be found in spreadsheet examples.xls and on the forced air tab. For this example we will use the curve above plus the details for an actual design.

<u>Design Particulars</u>

The design steam pressure for this ACC is 6.5 kPa (1.92 in.Hg). Considering the design steam quality (90) and latent heat, the heat removed per unit mass of steam at design conditions is 2170 kJ/kg (933 BTU/lbm). The design steam flow is 120 kg/s (952,397 lbm/hr). The design heat load is then 2.6×10^5 kW (8.88×10^8 BTU/hr). The design log-mean temperature difference (between the steam and air) is 20°C (36°F). The net heat exchange surface area is 461,189 m² (4,964,198 ft²). The design overall heat transfer coefficient, U, is then 28.2 W/m²/°C (4.97 BTU/hr/ft²/°F). This corresponds to an air velocity of 3.26 m/s (10.7 ft/sec). These details may be found on the Example15 tab of examples.xls and are summarized in the table on the following page.

Example 15. Calculations

	A	B	C	D	E	F
1	**Air-Cooled Steam Surface Condenser**					
2	Sample Calculations		SI Units		English Units	
3	INPUTS	symbol	units	value	units	value
4	steam pressure	Psat	kPa	6.50	in.Hg	1.92
5	steam temperature	Tsat	°C	37.6	°F	100
6	latent heat	hfg	kJ/kg	2411	BTU/lbm	1036
7	quality	x	-	90.0%	-	90.0%
8	steam flow	m	kg/s	120	lbm/hr	952,397
9	design LMTD	LMTD	°C	20	°F	36
10	design air temp.	Tdb	°C	10	°F	50
11	surface area	A	m²	461,189	ft²	4,964,198
12	CALCULATIONS					
13	unit heat transfer	q	kJ/kg	2170	BTU/lbm	933
14	heat transfer	Q	kW	2.60E+05	BTU/hr	8.88E+08
15	design conductance	U	W/m²/°C	28.2	BTU/hr/ft²/°F	4.97
16	design air velocity	V	m/s	3.26	ft/s	10.7
17	user inputs in blue					
18	calculations in orange					

Performance Curves

While the design point information and calculations are essential, performance curves (for prediction and evaluation) are also very important for this and most other heat exchangers. More discussion on that subject can be found elsewhere.[23] Operating points are obtained by solving the equations within the spreadsheet above at different conditions. These equations are nonlinear and so require iterative solution. The simplest approach is a binary search.[24] This approach is implemented in the spreadsheet in a VBA macro. The algorithm begins with a lower and upper limit on the estimate of the steam temperature required to achieve the total heat transfer and adjust the target up and down so as to match the log-mean temperature difference.

This ACC calculation macro accounts for the fraction of design steam flow (varying from 60% to 120%) and the fraction of the design air flow (varying from 60% to 120%) provided by the fan operation. The inlet air temperature and percentage of steam flow are varied to form the table on the following page.

[23] Benton, D. J., *Heat Exchangers: Performance Prediction & Evaluation*, ISBN-9781973589327, Amazon, 2017.

[24] Benton, D. J., *Nonlinear Equations: Numerical Methods for Solving*, ISBN-9781717767318, Amazon, 2018.

Calculated ACC Performance

air	steam pressure [kPa]				
Tdb	steamflow (percent of design)				
°C	80%	90%	100%	110%	120%
4.4	3.52	4.13	4.82	5.61	6.50
7.2	4.13	4.82	5.61	6.50	7.51
10.0	4.82	5.61	6.51	7.52	8.65
12.8	5.61	6.51	7.52	8.66	9.94
15.6	6.52	7.53	8.68	9.96	11.39
18.3	7.54	8.69	9.98	11.42	13.02
21.1	8.70	10.00	11.44	13.05	14.84
23.9	10.01	11.47	13.08	14.88	16.88
26.7	11.49	13.12	14.93	16.93	19.15
29.4	13.15	14.97	16.98	19.21	21.67
32.2	15.01	17.04	19.28	21.75	24.46
35.0	17.09	19.34	21.83	24.56	27.56
37.8	19.41	21.91	24.66	27.68	30.99

ACC Performance - SI Units

which results are also illustrated by these curves:

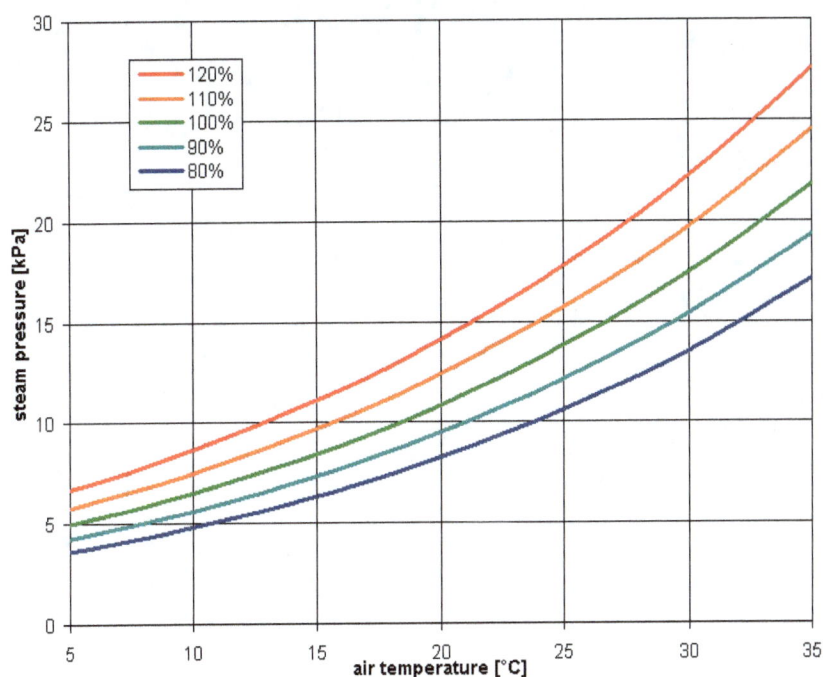

Figure 22. Calculated ACC Performance Curves

Example 16. E-Shell (Single-Pass) Liquid-Liquid

Shell-and-tube heat exchangers are used quite often, as they are efficient and easy to maintain (at least when compared to many other designs.) TEMA[25] is an excellent source for information about these.[26] There are several types. The first we will consider is called an E-Shell arrangement with one tube pass:

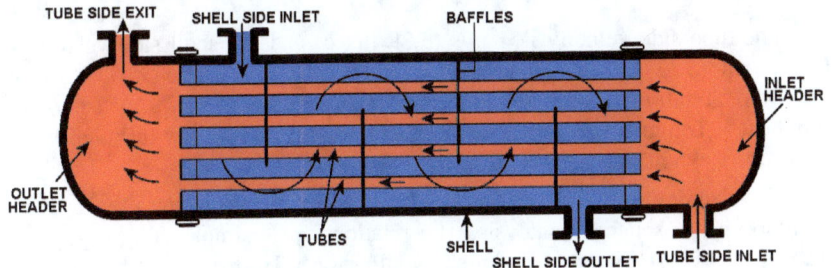

Figure 23. E-Shell Heat Exchanger

The hot fluid often flows through the tubes with the cold fluid in the shell but this is not always the case. While heat loss to or grain from the surroundings is a concern, pressure is usually the primary factor influencing this choice. The high-pressure fluid is almost always in the tubes. Avoiding leaks and reduced material costs must be considered for an optimal design.

Achieving a turbulent Reynolds number inside the tubes is rarely a problem; however, the shell must be designed to assure turbulence. To reduce the shell side cross-sectional flow area, baffles are added—as many as are necessary to assure turbulent convection but not so many as to cause significant separation zones.

For this example we will assume a tube length of 6 m (19.7 ft), a shell inside diameter of 1.25 m (4.1 ft), with 1200 tubes having an inside diameter of 1.5 cm (0.59 in) and outside diameter of 1.6 cm (0.63 in). The tube pitch is 3 cm (1.18 in) and there are 6 baffles. The heat transfer surface area is 362 m² (3896 ft²). The tube side flow area is 0.212 m² (2.28 ft²). The shell side flow area is 0.495 m² (5.31 ft²). The thermal conductivity of the tube material is 17.3 W/m/°C (10 BTU/hr/ft/°F).

The mass flow rates are 756,000 kg/hr (1,666,695 lbm/hr) on the tube side and 1,674,000 kg/hr (3,690,538 lbm/hr) on the shell side. The tube side fluid density is 997 kg/m³ (62.24 lbm/ft³), specific heat 4.18 kJ/kg/°C (0.998 BTU/lbm/°F), thermal conductivity 0.612 W/m/°C (0.354 BTU/hr/ft/°F), and dynamic viscosity 0.882 cP (2.13 lbm/ft/hr), making the tube side Prandtl number 6.02. The shell side fluid density is 987 kg/m³ (61.62 lbm/ft³), specific

---

[25] TEMA stands for Tubular Exchangers Manufacturers Association, a group of leading shell-and-tube designers and fabricators.

[26] https://tema.org/

heat 4.18 kJ/kg/°C (0.998 BTU/lbm/°F), thermal conductivity 0.641 W/m/°C (0.371 BTU/hr/ft/°F), and dynamic viscosity 0.559 cP (0.371 BTU/hr/ft/°F), making the Prandtl number 3.64.

$$G = \frac{\dot{m}}{A}$$

(16.1)

The tube side velocity is 1 m/s (3.29 ft/sec) and the shell side velocity is 0.954 m/s (3.13 ft/sec). The tube side mass flux, G, is 3.57×10^6 kg/hr/m² (7.30×10^6 lbm/hr/ft²). The shell side mass flux is 3.39×10^6 kg/hr/m² (6.94×10^6 lbm/hr/ft²). The tube side Reynolds number is then 17,012 (fully turbulent). The shell side Reynolds number is 26,956.

The Fanning (rather than Darcy) friction factor is most often used in shell-and-tube heat exchanger analysis. The relationship is simple $f_{Darcy}=4f_{Fanning}$. The Moody chart (Figure 14) can show either one. To distinguish between the two, the Darcy version will show $f=64/Re$ beside the laminar portion at the left side of the chart and the Fanning version will show f=16/Re. There is a function in the spreadsheet (examples.xls, see ff(Re)) providing an approximation of the fully-rough curve (Equation 16.2). The tube side friction factor using this formula is 0.00682 and the shell side friction factor is 0.00607. This and the Reynolds numbers provide a comparison of balance between the tube side and shell side.

$$f = \frac{16}{Re} \quad Re < 2000$$

(16.2)

$$f = \frac{1}{\left(1.58\log(Re)-3.28\right)^2} \quad Re > 8000$$

(16.3)

For Reynolds numbers between 2000 and 8000, linear interpolation is used. A similar empirical relationship is used for the Nusselt number:

$$Nu = 4.36 \quad Re < 2000$$

(16.4)

$$Nu = \frac{\left(\frac{f}{2}\right)Re\,Pr}{\left(1.07+12.7\sqrt{\frac{f}{2}}\right)\left(Pr^{\frac{2}{3}}-1\right)} \quad Re > 8000$$

(16.5)

As before, for Reynolds numbers between 2000 and 8000, linear interpolation is used. Using this piecewise correlation we get a tube side Nusselt number of 126 and a shell side of 147. Using the Nusselt number and thermal conductivity we gat a tube side (inside) heat transfer coefficient of 5121 W/m²/°C (902 BTU/hr/ft²/°F) and a shell side (outside) heat transfer coefficient of 5891 W/m²/°C (1038 BTU/hr/ft²/°F). Using Equation 11.1 we get a tube wall heat transfer coefficient (reciprocal of resistance) of 33,507 W/m²/°C (5,905 BTU/hr/ft²/°F).

When making calculations for heat exchangers of this type it is important to consider fouling, as these are essential components of an industrial system. If they fail, the entire system may also. There are numerous sources for fouling factors, which should consider fluids, materials, operating conditions, and frequency of maintenance. In this case we will assume a fouling conductance (reciprocal of resistance) of 2500 W/m²/°C (440 BTU/hr/ft²/°F). Fouling simply adds another term to Equation 11.2, by which we obtain an overall conductance, U, of 1280 W/m²/°C (218 BTU/hr/ft²/°F). Multiplying by the heat transfer surface are yields 463,395 W/°C (49,393 BTU/hr/°F).

LMTD vs. NTU-ε

While we could use the LMTD method for this heat exchanger (E-Shell), this is problematic for the other six TEMA designs. It is far easier to use the NTU-ε method, selecting the formula for the specific geometry. *Heat Transfer* by Lindon C. Thomas is an excellent source for these formula (and many more) plus detailed derivations and discussions.[27] To do this we must first define the heat capacity ratio, R:

$$R = \frac{C_{min}}{C_{max}} = \frac{\min(\dot{m}_C C_C, \dot{m}_H C_H)}{\max(\dot{m}_C C_C, \dot{m}_H C_H)} \qquad (16.6)$$

which for this problem is 0.452. The maximum possible heat transfer, Q_{max}, is:

$$Q_{max} = C_{min}(T_{hot,in} - T_{cold,in}) \qquad (16.7)$$

The tube (cold side) inlet temperature is 20°C (68°F) and the shell (hot side) inlet temperature is 50°C (122°F). The maximum heat transfer is then 2.63×10^4 kW (8.99×10^7 BTU/hr).

For this problem The actual heat transfer, Q, is:

$$Q = C_{hot}(T_{hot,in} - T_{hot,out}) = C_{cold}(T_{cold,out} - T_{cold,in}) \qquad (16.8)$$

The effectiveness, ε, is defined:

$$\varepsilon = \frac{Q}{Q_{max}} \qquad (16.9)$$

The number of transfer units, *NTU*, is defined:

---

[27] Thomas, L. C., Heat Transfer: Professional Version, Prentice Hall (1993) and Capstone Publishing (1999), which may be found at Amazon and other book sellers
https://www.amazon.com/Lindon-C.-Thomas/e/B001HOVPI4

$$NTU = \frac{UA}{C_{min}}$$

(16.10)

In this case NTU equals 0.528. For a counter-flow heat exchanger the effectiveness may be calculated thus:

$$\varepsilon = \frac{1 - \exp[-NTU(1-R)]}{1 - R\exp[-NTU(1-R)]}$$

(16.11)

For this example the effectiveness is equal to 0.38. The actual heat transfer is then 1×10^4 kW (3.33×10^7 BTU/hr). The tube side outlet temperature is 31.4°C (88°F) and the shell side outlet temperature is 44.9°C (113°F). The log-mean temperature difference (which we don't use) is 21.6°C (39.2°F). Details can be found in exaples.xls on the Example16 tab.

Example 16. Inputs

	A	B	C	D	E	F
1		Liquid-Liquid E-Shell (1-1 Pass)				
2	Sample Calculations		SI Units		English Units	
3	INPUTS	symbol	units	value	units	value
4	shell inside diameter	D_s	m	1.25	ft	4.10
5	tube length	L	m	6.0	ft	19.7
6	tube inside dia.	D_I	cm	1.5	in	0.591
7	tube outside dia.	D_o	cm	1.6	in	0.630
8	tube pitch	t_P	cm	3.0	in	1.181
9	baffles	N_B	-	6	-	6
10	tubes	N_T	-	1200	-	1200
11	tube material conduct.	k_T	W/m/°C	17.3	BTU/hr/ft/°F	10.0
12	tube side inlet temp.	$T_{C\,In}$	°C	20.0	°F	68.0
13	tube side density	ρ_T	kg/m³	997	lbm/ft³	62.2
14	tube side specific heat	C_T	kJ/kg/°C	4.18	BTU/lbm/°F	1.00
15	tube side dyn. visco.	μ_T	cP	0.882	lbm/ft/hr	2.13
16	tube side thermal cond.	k_T	W/m/°C	0.612	BTU/hr/ft/°F	0.354
17	tube side mass flow	m_T	kg/hr	756,000	lbm/hr	1,666,695
18	shell side inlet temp.	$T_{H\,In}$	°C	50.0	°F	122.0
19	shell side density	ρ_S	kg/m³	987	lbm/ft³	61.6
20	shell side specific heat	C_S	kJ/kg/°C	4.18	BTU/lbm/°F	1.00
21	shell side dyn. visco.	μ_S	cP	0.559	lbm/ft/hr	1.35
22	shell side thermal cond.	k_S	W/m/°C	0.641	BTU/hr/ft/°F	0.371
23	shell side mass flow	m_S	kg/hr	1,674,000	lbm/hr	3,690,538

Example 16. Calculations

	A	B	C	D	E	F
24	CALCULATIONS					
25	tube side flow area	A_T	m²	0.212	ft²	2.28
26	tube side velocity	V_T	m/s	1.00	ft/sec	3.29
27	tube side mass flux	G_T	kg/hr/m²	3.57E+06	lbm/hr/ft²	7.30E+05
28	tube side Reynolds	Re_T	-	17,012	-	17,012
29	tube side friction factor	f_T	-	0.00682	-	0.00682
30	tube side Prandtl	Pr_T	-	6.03	-	6.02
31	tube side Nusselt	Nu_T	-	126	-	125
32	tube side ht. tr.	h_T	W/m²/°C	5,121	BTU/hr/ft²/°F	902
33	shell side flow area	A_S	m²	0.494	ft²	5.31
34	shell side velocity	V_S	m/s	0.954	ft/sec	3.13
35	shell side mass flux	G_S	kg/hr/m²	3.39E+06	lbm/hr/ft²	6.94E+05
36	shell side Reynolds	Re_S	-	26,956	-	26,956
37	shell side friction factor	f_S	-	0.00607	-	0.00607
38	shell side Prandtl	Pr_S	-	3.65	-	3.64
39	shell side Nusselt	Nu_T	-	147	-	147
40	shell side ht. tr.	h_S	W/m²/°C	5,891	BTU/hr/ft²/°F	1,038
41	surface area	A	m²	362	ft²	3,896
42	tube thickness	t_G	cm	0.0500	in	0.0197
43	tube wall ht. tr. co.	h_W	W/m²/°C	335,071	BTU/hr/ft²/°F	5,905
44	fouling	h_F	W/m²/°C	2,500	BTU/hr/ft²/°F	440
45	overall ht. tr. co.	U	W/m²/°C	1,280	BTU/hr/ft²/°F	218
46	overall conductance	UA	W/°C	463,395	BTU/hr/°F	849,393
47	tube side heat capacity	$m_T C_T$	kJ/hr/°C	3.16E+06	BTU/hr/°F	1.66E+06
48	shell side heat capacity	$m_S C_S$	kJ/hr/°C	7.00E+06	BTU/hr/°F	3.69E+06
49	minimum heat capacity	mC_{min}	kJ/hr/°C	3.16E+06	BTU/hr/°F	1.66E+06
50	maximum heat capacity	mC_{max}	kJ/hr/°C	7.00E+06	BTU/hr/°F	3.69E+06
51	maximum heat transfer	Qmax	kW	2.63E+04	BTU/hr	8.99E+07
52	heat capacity ratio	R	-	0.452	-	0.452
53	number of transfer units	NTU	-	0.528	-	0.510
54	effectiveness	e	-	0.380	-	0.371
55	actual heat transfer	Q	kW	1.00E+04	BTU/hr	3.33E+07
56	tube side outlet temp.	$T_{c,out}$	°C	31.4	°F	88.0
57	shell side outlet temp.	$T_{H,out}$	°C	44.9	°F	113.0
58	log-mean temp. diff.	LMTD	°C	21.6	°F	39.2

Example 17. E-Shell (Two-Pass) Liquid-Liquid

E-Shell heat exchangers often have U-shaped tubes so that there is only one tube sheet and the ends are accessed through a single end, which can be unbolted. This configuration is often called an E-Shell1-2 arrangement:

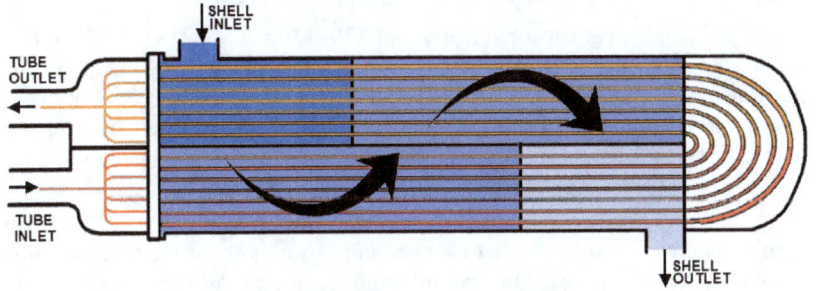

Figure 24. E-Shell 1-2 Arrangement

As these are often used for cooling, the hot fluid enters at the bottom and flows along and upward through the tubes. The cold fluid enters at the top near the hot side outlet so as to achieve the optimal terminal temperature difference and closest approach.

For this example there are 400 tubes of 4 m (13.1 ft) in length with 1.66 cm (0.634 in) inside diameter and 1.91 cm (0.752 in) outside diameter. The shell inside diameter is 0.4 m (1.31 ft) and there are 4 baffles. The heat exchange area is 96 m² (1033 ft²). The tube side flow area is 0.0433 m² (0.465 ft²). The shell side flow area is 0.0276 m² (0.297 ft²). The tube side mass flow rate is 200,000 kg/hr (440,925 lbm/hr) and the shell side flow rate is 100,000 kg/hr (220,462 lbm/hr). The tube side density is 700 kg/m³ (43.7 lbm/ft³) and the shell side density is 999 kg/m³ (62.4 lbm/ft³).

The tube side mass flux, G, is 4.62×10^6 kg/hr/m² (9.46×10^5 lbm/hr/ft²) and the shell side mass flux is 5.06×10^5 kg/hr/m² (1.04×10^5 lbm/hr/ft²). The tube side velocity is 1.83 m/s (6.02 ft/sec) and the shell side velocity is 1.01 m/s (3.31 ft/sec). The tube side dynamic viscosity is 1 cP (2.42 lbm/ft/hr) and the shell side dynamic viscosity is 0.5 cP (1.21 lbm/ft/hr). From these we calculate a tube side Reynolds number of 21,306 and a shell side of 38,488. Using the same formulas as in Example 16, the tube side friction factor is 0.00643 and the shell side 0.00557.

The thermal conductivity of the tube side fluid is 0.25 W/m/°C (0.145 BTU/hr/ft/°F) and the shell side is 0.5 W/m/°C (0.289 BTU/hr/ft/°F), making the tube side Prandtl number 8 and the shell side 4. Using the same formula as in Example 16, the tube side Nusselt number is 170 and the shell side 203. The turbulent convective heat transfer inside the tubes is then 2557 W/m²/°C (450 BTU/hr/ft²/°F) and the shell side is 5322 W/m²/°C (937 BTU/hr/ft²/°F).

The tube wall thickness is 0.125 cm (0.0492 in), making the tube wall conductance (reciprocal of resistance) 74,642 W/m²/°C (1,315 BTU/hr/ft²/°F). We will again assume a fouling conductance of 2,500 W/m²/°C (440 BTU/hr/ft²/°F), resulting in an overall conductance, U, of 953 W/m²/°C (151 BTU/hr/ft²/°F) and UA equal 91,458 W/°C (155,553 BTU/hr/°F).

The tube side heat capacity, $m_T C_T$, is 8×10^5 kJ/hr/°C (4.21×10^5 BTU/hr/°F) and the shell side, $m_S C_S$, is 4×10^5 kJ/hr/°C (2.11×10^5 BTU/hr/°F). The maximum heat transfer, Q_{max}, is 6.67×10^3 kW (2.27×10^7 BTU/hr). The heat capacity ratio, R, is 0.5 and the number of transfer units, NTU, is 0.738 (unitless). The formula for effectiveness is considerably more complicated for this arrangement, which is why the NTU-ε method is preferred to the LMTD in such cases. This arrangement is called an E-Shell (1-2) with the shell side fluid "mixed"; that is, the fluid on the shell side is able to flow along as well as over the tubes. In order for the fluid to be "unmixed" this would require some sort of partitions:

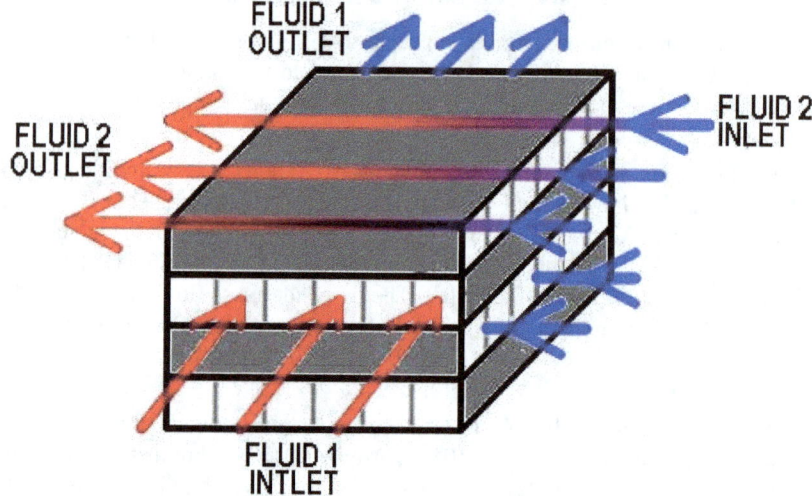

Figure 25. Heat Exchanger with Unmixed Streams

For this arrangement we calculate P:[28]

$$\varepsilon = \frac{4}{2(1+R) + \Gamma \coth\left(\frac{\Gamma NTU}{4}\right) + \tanh\left(\frac{NTU}{4}\right)}$$

(17.1)

where Γ is:

$$\Gamma = \sqrt{4R^2 + 1}$$

(17.2)

---

[28] See *Heat Transfer* by L. C. Thomas Appendix N

The effectiveness is 0.459 (unitless) and the heat transfer is 3057 kW (1.04×10^7 BTU/hr). The tube side outlet temperature is 66.2°C (151°F) and the shell side outlet temperature is 47.5°C (118°F). The LMTD is 39°C (70°F). The inputs are shown in the following table:

Example 17. Inputs

	A	B	C	D	E	F
1	Liquid-Liquid E-Shell (1-2 Pass)					
2	Sample Calculations		SI Units		English Units	
3	INPUTS	symbol	units	value	units	value
4	shell inside diameter	D_s	m	0.50	ft	1.64
5	tube length	L	m	4.0	ft	13.1
6	tube inside dia.	D_I	cm	1.66	in	0.654
7	tube outside dia.	D_O	cm	1.91	in	0.752
8	tube pitch	t_p	cm	2.1	in	0.827
9	tube passes	N_P	-	2	-	2
10	baffles	N_B	-	4	-	4
11	tubes	N_T	-	400	-	400
12	tube material conduct.	k_T	W/m/°C	10.0	BTU/hr/ft/°F	5.78
13	tube side inlet temp.	$T_{C In}$	°C	80.0	°F	176
14	tube side density	ρ_T	kg/m³	700	lbm/ft³	43.7
15	tube side specific heat	C_T	kJ/kg/°C	2.00	BTU/lbm/°F	0.48
16	tube side dyn. visco.	μ_T	cP	1.0	lbm/ft/hr	2.42
17	tube side thermal cond.	k_T	W/m/°C	0.25	BTU/hr/ft/°F	0.145
18	tube side mass flow	m_T	kg/hr	200,000	lbm/hr	440,925
19	shell side inlet temp.	$T_{H In}$	°C	20	°F	68
20	shell side density	ρ_S	kg/m³	999	lbm/ft³	62.4
21	shell side specific heat	C_S	kJ/kg/°C	4.00	BTU/lbm/°F	0.96
22	shell side dyn. visco.	μ_S	cP	0.5	lbm/ft/hr	1.21
23	shell side thermal cond.	k_S	W/m/°C	0.5	BTU/hr/ft/°F	0.289
24	shell side mass flow	m_S	kg/hr	100,000	lbm/hr	220,462

The calculations are shown below:

Example 17. Calculations

	A	B	C	D	E	F
25	CALCULATIONS					
26	tube side flow area	A_T	m²	0.0433	ft²	0.466
27	tube side velocity	V_T	m/s	1.83	ft/sec	6.02
28	tube side mass flux	G_T	kg/hr/m²	4.62E+06	lbm/hr/ft²	9.46E+05
29	tube side Reynolds	Re_T	-	21,306	-	21,306
30	tube side friction factor	f_T	-	0.00643	-	0.00643
31	tube side Prandtl	Pr_T	-	8.00	-	7.99
32	tube side Nusselt	Nu_T	-	170	-	170
33	tube side ht. tr.	h_T	W/m²/°C	2,556	BTU/hr/ft²/°F	450
34	shell side flow area	A_S	m²	0.0276	ft²	0.297
35	shell side velocity	V_S	m/s	1.009	ft/sec	3.31
36	shell side mass flux	G_S	kg/hr/m²	3.63E+06	lbm/hr/ft²	7.43E+05
37	shell side Reynolds	Re_S	-	38,489	-	38,489
38	shell side friction factor	f_S	-	0.00557	-	0.00557
39	shell side Prandtl	Pr_S	-	4.00	-	4.00
40	shell side Nusselt	Nu_T	-	205	-	205
41	shell side ht. tr.	h_S	W/m²/°C	5,372	BTU/hr/ft²/°F	946
42	surface area	A	m²	96	ft²	1,033
43	tube thickness	t_Θ	cm	0.125	in	0.0492
44	tube wall ht. tr. co.	h_W	W/m²/°C	7,464	BTU/hr/ft²/°F	1,315
45	fouling	h_f	W/m²/°C	2,500	BTU/hr/ft²/°F	440
46	overall ht. tr. co.	U	W/m²/°C	854	BTU/hr/ft²/°F	151
47	overall conductance	UA	W/°C	82,035	BTU/hr/°F	155,553
48	tube side heat capacity	$m_T C_T$	kJ/hr/°C	8.00E+05	BTU/hr/°F	4.21E+05
49	shell side heat capacity	$m_S C_S$	kJ/hr/°C	4.00E+05	BTU/hr/°F	2.11E+05
50	minimum heat capacity	mC_{min}	kJ/hr/°C	4.00E+05	BTU/hr/°F	2.11E+05
51	maximum heat capacity	mC_{max}	kJ/hr/°C	8.00E+05	BTU/hr/°F	4.21E+05
52	maximum heat transfer	Qmax	kW	6.67E+03	BTU/hr	2.27E+07
53	heat capacity ratio	R	-	0.500	-	0.500
54	number of transfer units	NTU	-	0.738	-	0.739
55	gamma	Γ	-	1.414	-	1.414
56	effectiveness	ε	-	0.459	-	0.459
57	actual heat transfer	Q	kW	3,057	BTU/hr	1.04E+07
58	tube side outlet temp.	$T_{c,out}$	°C	66.2	°F	151
59	shell side outlet temp.	$T_{H,out}$	°C	47.5	°F	118
60	log-mean temp. diff.	LMTD	°C	39.0	°F	70.1

The temperature variation through the tubes, along the tube surface, and through the shell are shown in this next figure:

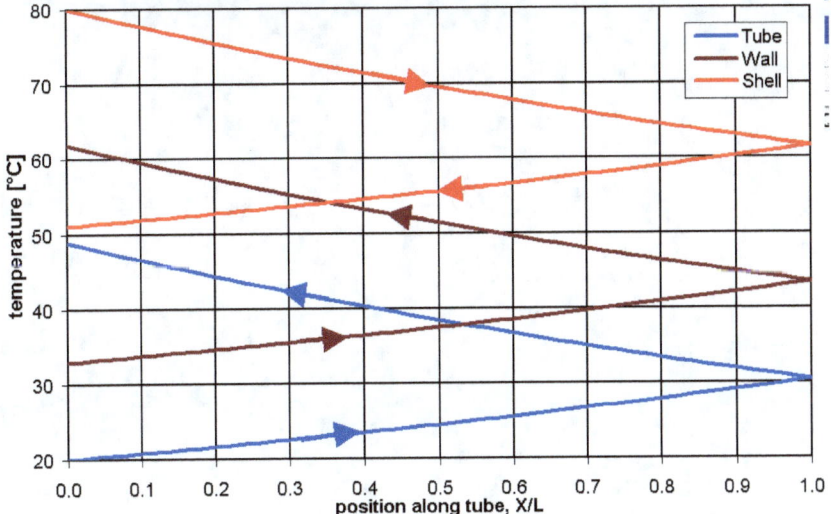

Figure 26. Example 17 Temperature Profiles

Example 18. HRSG Economizer

A Heat Recovery Steam Generator (HRSG) is a specific type of compound heat exchanger attached to the exhaust of a combustion gas turbine so as to *recover* (i.e., make use of) the waste heat. These devices create superheated steam, which flows through a steam turbine connected to a generator to make electricity. The combine apparatus has greater efficiency than either of the two parts (gas turbine + steam turbine) by itself. A typical single-pressure HRGS is illustrated in this next figure:

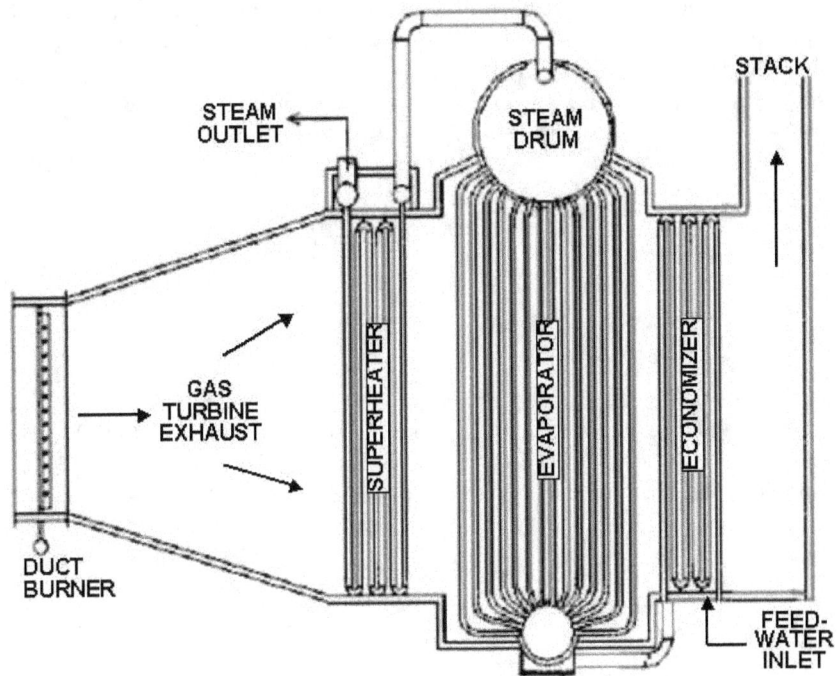

Figure 27. Typical Single-Pressure HRSG

Compressed water enters from the right, flows through the economizer, where it approaches—but does not reach—the saturated liquid state. The hot water then enters the evaporator, where it boils. Steam vapor bubbles rise through the vertical tubes and collect in the steam drum. Saturated vapor leaves the steam drum and flows through the superheater. Superheated steam then flows to and through the steam turbine. In this chapter we will consider the economizer.

Water enters the economizer at 8274 kPa (1200 psia) and 44.4°C (112°F). The water specific heat is 4.41 kJ/kg/°C (1.05 BTU/lbm/°F) and the flow is 31.9 kg/s (253,215 lbm/hr). Exhaust enters the economizer at approximately atmospheric pressure and 314°C (598°F). The exhaust specific heat is 1.08

kJ/kg/°C (0.258 BTU/lbm/°F) and the flow is 317.3 kg/s (2,518,234 lbm/hr). The heat exchange surface area is 7,498 m² (80,705 ft²) and the overall conductance, U, is 45.4 W/m²/°C (8 BTU/hr/ft²/°F). Note that while we could apply similar correlations for inside and outside finned tubes here, a constant value is more commonly used for design calculations. overall conductance. The product, UA, is then 341 kW/°C (645,640 BTU/hr/°F).

The water side heat capacity, $m_W C_W$, is 141 kW/°C (266,635 BTU/hr/°F) and the exhaust side heat capacity, $m_G C_G$, is 343 kW/°C (649,704 BTU/hr/°F). The heat capacity ratio, R, is then 0.41 and the number of transfer units, NTU, is equal to 2.42. The NTU-ε method is most often used in these cases, as the elements of a HRSG are a crossflow arrangement with multiple tube passes and the adjustment for tube passes is relatively simple. Here we will use the following:

$$\varepsilon = \frac{2}{1 + R + \beta\sqrt{1 + R^2}}$$
$$\Gamma = NTU\sqrt{1 + R^2} \qquad (18.1)$$
$$\beta = \frac{1 + e^{-\Gamma}}{1 - e^{-\Gamma}}$$

from which we calculate an effectiveness, ε, of 0.751, a water outlet temperature of 247°C (477°F), and a exhaust gas outlet temperature of 231°C (448°F). The LMTD is 117°C (210°F). The heat transfer is 28,527 kW (1.3×10^8 BTU/hr). Complete details can be found on the Example18 tab of examples.xls.

Example 18. Details

	A	B	C	D	E	F
1			HRSG Economizer			
2	Sample Calculations		SI Units		English Units	
3	INPUTS	symbol	units	value	units	value
4	water inlet pressure	P_W	kPa	8274	psia	1200
5	water inlet temperature	T_{W1}	°C	44.4	°F	112
6	water specific heat	C_W	kJ/kg/°C	4.41	BTU/lbm/°F	1.05
7	water flow	m_W	kg/s	31.9	lbm/hr	253,215
8	exh. inlet temperature	T_{G1}	°C	314	°F	598
9	exh. specific heat	C_G	kJ/kg/°C	1.08	BTU/lbm/°F	0.258
10	exh. flow	m_G	kg/s	317.3	lbm/hr	2,518,234
11	surface area	A	m²	7,498	ft²	80,705
12	overall conductance	U	W/m²/°C	45.4	BTU/hr/ft²/°F	8.00
13	CALCULATIONS					
14	overall conductance	UA	kW/°C	341	BTU/hr/°F	645,640
15	water heat capacity	$m_T C_T$	kW/°C	141	BTU/hr/°F	266,635
16	exh. heat capacity	$m_B C_B$	kW/°C	343	BTU/hr/°F	649,704
17	minimum heat capacity	mC_{min}	kW/°C	141	BTU/hr/°F	266,635
18	maximum heat capacity	mC_{max}	kW/°C	343	BTU/hr/°F	649,704
19	maximum heat transfer	Qmax	kW	37,963	BTU/hr	1.30E+08
20	heat capacity ratio	R	-	0.410	-	0.410
21	number of transfer units	NTU	-	2.42	-	2.42
22	gamma	Γ	-	2.617	-	2.617
23	beta	β	-	1.157	-	1.157
24	effectiveness	ε	-	0.751	-	0.751
25	actual heat transfer	Q	kW	28,527	BTU/hr	9.73E+07
26	water outlet temp.	Tw_{out}	°C	247	°F	477
27	exh. outlet temp.	$T_{G,out}$	°C	231	°F	448
28	log-mean temp. diff.	LMTD	°C	117	°F	210
29			user inputs in blue			
30			calculations in orange			

Example 19. HRSG Evaporator

The evaporator section (see Figure 27) consists of vertical tubes. Boiling occurs within the tubes and the steam rises to accumulate in the steam drum. Even though this is a crossflow arrangement, the steam side is isothermal (constant at the saturation temperature), as there is very little pressure drop in the tubes; therefore, we can use the LMTD to calculate the heat transfer.

We will assume that the water enters at the saturated liquid state (no subcooling); therefore, the entering and exiting temperature is controlled by the pressure, which is 7929 kPa (1150 psia). The saturation temperature of steam at this pressure is 294°C (572°). The water mass flow rate is 31.9 kg/s (253,215 lbm/hr). The exhaust gas enters at 455°C (851°F). The specific heat of exhaust gas at these conditions (close to ambient pressure) is 1.11 kJ/kg/°C (0.266 BTU/lbm/°F).

At this point we need steam properties to continue. The spreadsheet (examples.xls) contains a VBA implementation of the 1967 ASME steam tables. There are several functions, including one for the saturated liquid, h_F, and one for the saturated vapor, hG. Using these relationships we get an entering water enthalpy of 1313 kJ/kg (564 BTU/lbm) and an exiting enthalpy of 2760 kJ/kg (1187 BTU/lbm). The total heat transfer is then 46,174 kW (1.58×10^8 BTU/hr).

The exiting gas temperature can be calculated from the heat transfer, mass flow rate, and specific heat to obtain a value of 324°C (616°F). The LMTD is then 77.6°C (140°F). This heat exchanger must have a UA of 595 kW/°C (1.13×106 BTU/hr/°F).

Example 19. Details

	A	B	C	D	E	F
1				HRSG Evaporator		
2	Sample Calculations			SI Units		English Units
3	INPUTS	symbol	units	value	units	value
4	water inlet pressure	P_W	kPa	7929	psia	1150
5	water inlet temperature	T_{W1}	°C	294	°F	562
6	water flow	m_W	kg/s	31.9	lbm/hr	253,215
7	exh. inlet temperature	T_{G1}	°C	455	°F	851
8	exh. specific heat	C_G	kJ/kg/°C	1.11	BTU/lbm/°F	0.266
9	exh. flow	m_G	kg/s	317	lbm/hr	2,518,234
10	CALCULATIONS					
11	water inlet enthalpy	h_f	kJ/kg	1313	BTU/lbm	564
12	water exit enthalpy	h_G	kJ/kg	2760		1187
13	heat transfer	Q	kW	46,174	BTU/hr	1.58E+08
14	exh. outlet temp.	$T_{G,out}$	°C	324	°F	616
15	log-mean temp. diff.	LMTD	°C	77.6	°F	140
16	overall conductance	UA	kW/°C	595	BTU/hr/°F	1.13E+06

Example 20. HRSG Superheater

The superheater section (see Figure 27) also consists of vertical tubes (often U-shaped) with steam flowing vertically down and up. These are crossflow heat exchangers with the gas side (outside the tubes) fully mixed and the steam side (inside the tubes) not mixed; therefore, we will use the NTU-ε as before.

Steam enters at a pressure of 7929 kPa (1150 psia) and a temperature of 294°C (562°F). The specific heat of steam at this pressure is 3.16 kJ/kg/°C (0.755 BTU/lbm/°F). The hot gas turbine exhaust gas enters at a temperature of 509°C (949°F), having a specific heat of 1.14 kJ/kg/°C (0.272 BTU/lbm/°F), and a flow rate of 317 kg/s (2,518,235 lbm/hr).

The heat exchange surface area is 7373 m² (79,363 ft²) and we assume a conductance, U, of 45.4 W/m²/°C (8 BTU/ft²/°F). The overall conductance, UA, is then 335 kW/°C (634,904 BTU/hr/°F). The steam heat capacity, $m_S C_S$, is 101 kW/°C (191,177 BTU/hr/°F) and the gas stream heat capacity, $m_G C_G$, is 361 kW/°C (684,960 BTU/hr/°F). The maximum possible heat transfer, $Qmax$, us 21,689 kW (7.4×10⁷ BTU/hr).

The heat capacity ratio, R, is 0.279 (unitless) and the number of transfer units, NTU, is 3.23 (also unitless). Using Equation 18.1 we arrive at an effectiveness, ε, of 0.838 (unitless). The actual heat transfer, Q, is 18,183 kW (6.2×10⁷ BTU/hr). The steam outlet temperature is 475°C (886°F) and the gas outlet temperature is 459°C (858°F). The LMTD is 84°C (150°F). Details are listed on the next page.

Example 20. Details

	A	B	C	D	E	F
1		HRSG Superheater				
2	Sample Calculations		SI Units		English Units	
3	INPUTS	symbol	units	value	units	value
4	steam inlet pressure	P_W	kPa	7929	psia	1150
5	steam inlet temperature	T_{W1}	°C	294	°F	562
6	steam specific heat	C_W	kJ/kg/°C	3.16	BTU/lbm/°F	0.76
7	steam flow	m_W	kg/s	31.9	lbm/hr	253,215
8	exh. inlet temperature	T_{G1}	°C	509	°F	949
9	exh. specific heat	C_G	kJ/kg/°C	1.14	BTU/lbm/°F	0.272
10	exh. flow	m_G	kg/s	317	lbm/hr	2,518,234
11	surface area	A	m²	7,373	ft²	79,363
12	overall conductance	U	W/m²/°C	45.4	BTU/hr/ft²/°F	8.00
13	CALCULATIONS					
14	overall conductance	UA	kW/°C	335	BTU/hr/°F	634,904
15	steam heat capacity	$m_T C_T$	kW/°C	101	BTU/hr/°F	191,177
16	exh. heat capacity	$m_S C_S$	kW/°C	361	BTU/hr/°F	684,960
17	minimum heat capacity	mC_{min}	kW/°C	101	BTU/hr/°F	191,177
18	maximum heat capacity	mC_{max}	kW/°C	361	BTU/hr/°F	684,960
19	maximum heat transfer	Qmax	kW	21,689	BTU/hr	7.40E+07
20	heat capacity ratio	R	-	0.279	-	0.279
21	number of transfer units	NTU	-	3.32	-	3.32
22	gamma	Γ	-	3.448	-	3.448
23	beta	β	-	1.066	-	1.066
24	effectiveness	ε	-	0.838	-	0.838
25	actual heat transfer	Q	kW	18,183	BTU/hr	6.20E+07
26	steam outlet temp.	$Tw_{,out}$	°C	475	°F	886
27	exh. outlet temp.	$T_{G,out}$	°C	459	°F	858
28	log-mean temp. diff.	LMTD	°C	84	°F	150
29		user inputs in blue				
30		calculations in orange				

Example 21. Feedwater Heater

There are three factors that complicate heat exchanger analysis: varying properties, geometry, and phase change. Feedwater heaters are used in Rankine cycles (i.e., steam power plants) to heat the water from the condenser before it enters the boiler, which significantly improves the overall thermal efficiency for reasons discussed elsewhere. Feedwater heaters most often have three zones: de-superheating, condensing, and sub-cooling. In order to analyze this type of heat exchanger, we must consider these three zones separately, as shown below:

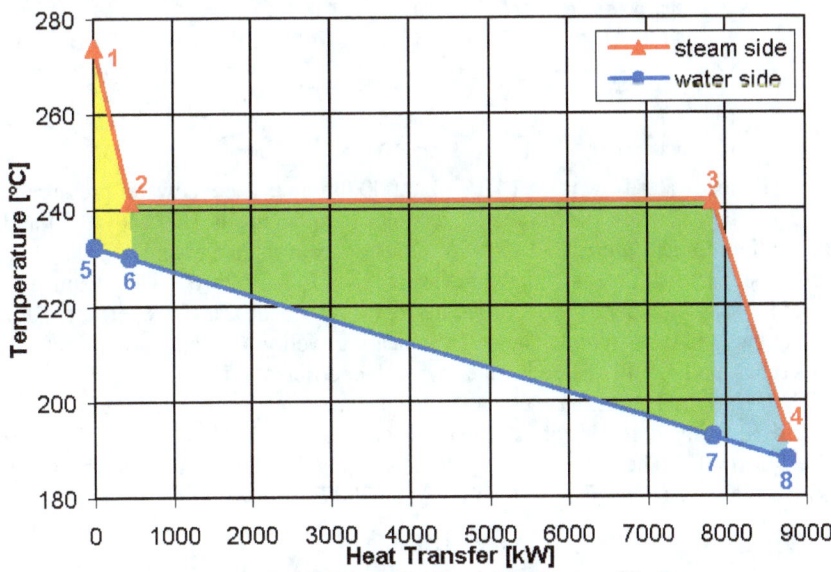

Figure 28. Four Zones Inside A Feedwater Heater

The yellow area represents the de-superheating zone, in which the steam is cooled from a superheated state to saturated vapor. The red process line has a sharp downward slope between points 1 and 2. The green area represents the condensing zone, in which the steam condenses on the outside of the tube bundle. The red process line is flat over this zone because the saturation temperature is constant. The cyan area represents the sub-cooling zone, in which the condensate is further cooled, as it approaches the inlet feedwater temperature. The red process line from point 3 to 4 has a downward slope that is not as steep as the line from point 1 to 2. The slope of the lines from 1 to 2 to 3 to 4 are inversely proportional to the specific heat times the mass flow rate. The specific heat of steam (vapor) is approximately half that of water (liquid), thus the difference. The specific heat at saturation is infinite ($\Delta h/\Delta T = \infty$ because $\Delta T = 0$), thus the flat segment of the process line (slope$=1/\infty=0$). We first consider the inputs or design parameters:

Example 21. Inputs

	A	B	C	D	E	F
1			**Feed Water Heater Example**			
2	Sample Calculations		SI Units		English Units	
3	INPUTS	symbol	units	value	units	value
4	FW flow	m_f	kg/s	**44.1**	lbm/hr	**350,000**
5	FW inlet pres.	P_{FI}	kPa	**15,168**	psia	**2200**
6	FW inlet temp.	T_{FI}	°C	**188**	°F	**370**
7	FW exit temp.	T_{FO}	°C	**232**	°F	**450**
8	stm. pressure	P_S	kPa	**3,447**	psia	**500**
9	steam in temp.	T_{SI}	°C	**274**	°F	**525**
10	steam exit temp.	T_{SO}	°C	**193**	°F	**380**
11	desuperheating	U_S	W/m²/°C	**568**	BTU/hr/ft²/°F	**100**
12	condensing	U_C	W/m²/°C	**3407**	BTU/hr/ft²/°F	**600**
13	subcooling	U_D	W/m²/°C	**1703**	BTU/hr/ft²/°F	**300**

The feedwater flow is 44.1 kg/s (350,000 lbm/hr). The feedwater pressure is 15,168 kPa (2200 psia). The feedwater inlet temperature is 188°C (370°F) and the outlet temperature is 232°C (450°F). The steam inlet pressure is 3,447 kPa (500 psia) and the steam inlet temperature is 274°C (525°F). The steam exit (drain cooler) temperature is 193°C (380°F). The heat transfer coefficient for each zone will be different. The de-superheating (yellow) zone has vapor outside the tubes and liquid inside with a typical coefficient of 568 W'/m²/°C (100 BTU/hr/ft²/°F). The next zone (green) has steam condensing outside tubes with liquid flowing within for a typical coefficient of 3407 W'/m²/°C (600 BTU/hr/ft²/°F). The final zone (cyan) has liquid inside and outside the tubes with a typical coefficient of 1703 W'/m²/°C (300 BTU/hr/ft²/°F).

Example 21. Table of Zones

G	H	I	J	K	L	M	N
				water side			
	SI Units				English Units		
T	P	H	Q	T	P	H	Q
°C	kPa	kJ/lg	kW	°F	psia	BTU/lbm	MBTU/hr
232	15,168	1003	0	450	2200	431	0.0
230	15,168	993	440	446	2200	427	1.5
193	15,168	826	7830	379	2200	355	26.7
188	15,168	804	8765	370	2200	346	29.9
				steam side			
	SI Units				English Units		
T	P	H	Q	T	P	H	Q
°C	kPa	kJ/lg	kW	°F	psia	BTU/lbm	MBTU/hr
274	3,447	2907	0	525	500	1250	0
242	3,447	2802	440	467	500	1205	1.5
242	3,447	1046	7830	467	500	450	26.7
193	3,447	823	8765	380	500	354	29.9

We begin this table by filling in the pressures and temperatures. T1 and T4 are listed in the inputs. T2 and T3 are equal to the saturation temperature, which is calculated from the pressure using the steam property functions embedded in the spreadsheet. T5 and T8 are listed in the inputs but T6 and T7 must be calculated from the enthalpies at these points. For each of the zones the heat transfer from the steam goes to the feedwater, as we are neglecting heat loss to the surroundings. For each zone we have the following relationship:

$$\dot{m}\Delta h_{FW} = \dot{Q}_{FW} = -\dot{Q}_{STEAM} = -\dot{m}\Delta h_{STEAM} \qquad (21.1)$$

The two intermediate feedwater temperatures (T5 and T6) can be calculated linearly from the respective enthalpies as the specific over the range is approximately constant. The enthalpies are required for these calculations and are shown in the following table:

Example 21. Properties and Calculations

	A	B	C	D	E	F
		symbol	units	value	units	value
14	CALCULATIONS					
15	FW inlet	h_{FI}	kJ/kg	804	BTU/lbm	346
16	FW exit	h_{FO}	kJ/kg	1003	BTU/lbm	431
17	heat transfer	Q	kW	8771	BTU/hr	2.99E+07
18	saturation temp.	T_{SAT}	°C	242	°F	467
19	stm. in enthalpy	h_{SI}	kJ/kg	2907	BTU/lbm	1250
20	stm. sat. vapor	h_G	kJ/kg	2802	BTU/lbm	1205
21	stm. sat. liquid	h_F	kJ/kg	1046	BTU/lbm	450
22	stm. exit enth.	h_D	kJ/kg	823	BTU/lbm	354
23	steam flow	m_S	kg/s	4.21	lbm/hr	33,412
24	FW enth at h_G	h_{FWG}	kJ/kg	993	BTU/lbm	427
25	FW temp at h_G	T_{FWG}	°C	230	°F	446
26	FW enth at h_F	h_{FWF}	kJ/kg	826	BTU/lbm	355
27	FW temp at h_F	T_{FWF}	°C	193	°F	379
28	desuperheating	LMTD	°C	23.6	°F	42.4
29	condensing	LMTD	°C	26.1	°F	46.9
30	subcooling	LMTD	°C	20.0	°F	36.0
31	desuperheating	A	m²	32.9	ft²	354
32	condensing	A	m²	83.2	ft²	896
33	subcooling	A	m²	27.4	ft²	296
34	total	A	m²	144	ft²	1,546

We can use the LMTD over each of the three zones but not over the entire zone. The NTU-ε can also be used on a zone-by-zone basis but there is no point and the calculations are more tedious. The three log-mean temperature differences are calculated using the VBA macro:

$$\Delta T_{des} = LMTD(T1 - T5, T2 - T6) \qquad (21.2)$$

$$\Delta T_{cond} = LMTD(T2 - T6, T3 - T7) \qquad (21.3)$$

$$\Delta T_{sub} = LMTD(T3 - T7, T4 - T8) \qquad (21.4)$$

We calculate the required area for each zone by:

$$A = \frac{\dot{Q}}{U\Delta T} \qquad (21.5)$$

Finally, we sum up the required areas from which we calculate the number of tubes, having beforehand selected a tube size, length, and material suitable for the application.

Example 22. Fogger

Sometimes (especially in dry climates) a fine water mist will be sprayed at the inlet of a combustion gas turbine to reduce the temperature. While this isn't technically a heat exchanger, it is a common device used to control temperature and solving it will provide an introduction to psychrometrics (moist air properties), which date back to the work of Goff & Gratch.[29] Their formulas have been implemented as VBA macros in the spreadsheet examples.xls

In order to use these formulas, we must define some terms unique to psychrometrics. The dry-bulb temperature is the common temperature of the air, historically measured with a dry mercury-in-glass thermometer. The wet-bulb temperature is measured by placing a moist "sock" on a thermometer (or thermocouple or (RTD) resistance temperature device) and passing air over it at a modest speed. The air speed is selected to achieve a Lewis number of unity, which is a mass transfer concept and beyond the scope of this text. The wet-bulb temperature is assumed to be equivalent to the adiabatic (zero heat transfer) saturation temperature (the temperature you would reach if you saturated the air with water vapor without transferring heat to or from the air).

The dew point temperature is the temperature to which the air must be cooled to initiate condensation. The dew point of perfectly dry air is undefined because no degree of cooling will result in condensation. As it turns out, the barometric pressure and dew point uniquely define the moisture content of air. While the dew point and wet-bulb are similar, they are not equal, except at the saturation point.

Relative humidity is the ratio of the actual to maximum water bearing capacity of the air, most often expressed as a percentage. If there is no moisture, the relative humidity is 0%. If the air is saturated (at the point of condensation), then the relative humidity is 100%. At a relative humidity of 100%, the dry-bulb, wet-bulb, and dew point are all equal. Note that percentages in Excel vary from zero to one, even though they may be displayed next to a % sign as zero to 100.

The humidity ratio, W, is an important part of psychrometric calculations. This is the mass of water per mass of dry air and is *not* equal to the relative humidity. One thing peculiar to psychrometrics is the definition of terms (such as density and enthalpy) per unit mass of dry air, not per total (moist) mass (air plus water vapor). Humidity ratio is *not* equal to the mass (or mole) fraction of water vapor in (moist) air.

We don't want to spray so much water that the engine aspirates rain, thus we select some target relative humidity, say 90%. We are not *cooling* the air, as might be the case with a chiller (i.e., an air conditioning unit). We are simply

---

[29] Goff, J. A. and Gratch, S., "Thermodynamic Properties of Moist Air," Heating, Piping & Air Conditioning, pp. 334-348, 1945.

introducing more water at roughly the same temperature as the ambient air. So this is not a dew-point process (cooling/condensing), it is more like a partial wet-bulb process. In fact, fogger performance is often reported as "achieving 90% of the wet-bulb".

How much fogging (mass flow rate of water) is needed can then be calculated. This is equal to the mass flow rate of dry air times the difference in humidity ratio at the target humidity minus the humidity ratio of the ambient.

$$\dot{m}_{FOGGER} = \dot{m}_{DRYAIR}\left(W_{TARGET} - W_{AMBIENT}\right) \qquad (22.1)$$

There is a function (fWdbrh) provided which returns the humidity ratio given the barometric pressure, dry-bulb temperature, and relative humidity. We can dig deeper into this calculation... As no heat is added or removed, the enthalpy before and after fogging will be the same. Psychrometric enthalpy is per unit mass of dry air. The flow of dry air does not change so the psychrometric enthalpy is also the same. In this case the same principle holds but we must always be mindful of this distinction (per unit total mass vs. per unit mass of dry air) when working with psychrometrics. The enthalpy (per unit mass of dry air) is given by:

$$h_{MA} = C_{PA}T + Wh_G \qquad (22.2)$$

where h_{MA} is the enthalpy of moist air per unit mass of dry air, C_{PA} is the specific heat of dry air, W is the humidity ratio (mass of water vapor per mass of dry air), and h_G is the enthalpy of saturated water vapor. Note that this last term is **not** h_{FG}, the latent heat of water. Errors exist in the literature—even in respected reference books—regarding this fact. Consider this fact: Introducing water vapor at the critical point (374°C or 704°F) will most certainly increase the enthalpy, yet the latent heat, h_{FG}, at the critical point is zero by definition.

Recalling that the humidity ratio, W, is a function of barometric pressure, dry-bulb temperature, and relative humidity, we can write the following:

$$C_{PA}T_{AMBIENT} + W_{AMBIENT}h_G = C_{PA}T_{FOGGED} + W_{FOGGED}h_G \qquad (22.3)$$

We can then write a VBA macro to solve this nonlinear equation. It is nonlinear because it is implicit in the final (fogged) temperature and W contains multiple nonlinear relationships (saturation pressure of water vapor and the correction factor representing the departure from ideal behavior). The macro uses a bisection search to find the fogged temperature such that the enthalpy before and after are equal:

```
Function Tfog(ByVal baro As Double, ByVal Tamb As
    Double, ByVal RHamb As Double, ByVal RHfog As Double)
    As Double
    Dim iter As Integer, h1 As Double, h2 As Double,
        T1 As Double, T2 As Double
    If (RHamb >= RHfog) Then
```

```
      Tfog = Tamb
      Exit Function
    End If
    h1 = fHtbrh(baro, Tamb, RHamb)
 'the wet-bulb is the lowest possible temperature
    T1 = fBdbrh(baro, Tamb, RHamb)
 'the ambient is the highest possible temperature
    T2 = Tamb
    For iter = 1 To 32
      Tfog = (T1 + T2) / 2
      h2 = fHtbrh(baro, Tfog, RHfog)
      If (h2 < h1) Then
        T1 = Tfog
      Else
        T2 = Tfog
      End If
    Next iter
  End Function
```

The following graph shows the results. The amount of water required can be calculated from the temperatures using Equation 22.1.

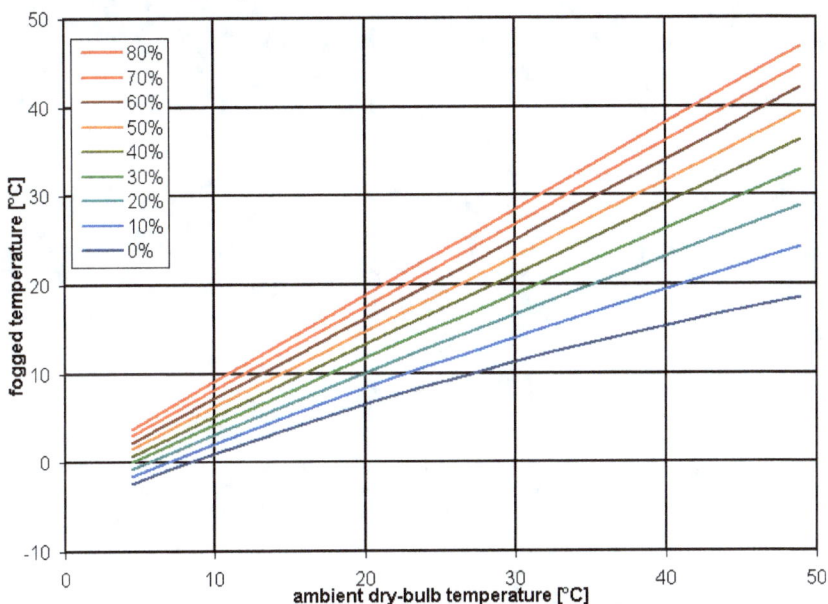

Figure 29. Combustion Gas Turbine Inlet Fogging Temperature

For more details on psychrometrics see Appendix A.

Example 23. Mechanical Draft Counterflow Cooling Tower

There are many designs for mechanical induced-draft evaporative cooling towers. In this chapter we will consider one common arrangement: rectangular counterflow. Induced-draft means that the fans draw air through the tower and are mounted at the top. Counterflow means that the water falls vertically downward and the air flows vertically upward through the packing or fill. Rectangular means that air flows in from all sides. There is typically a single fan per cell, though some designs use two. This type of tower is illustrated below:

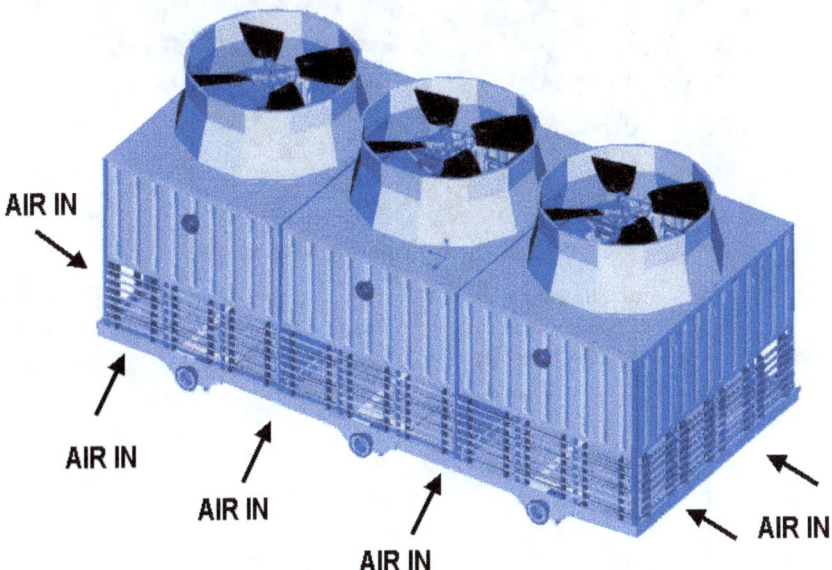

Figure 30. Typical Mechanical Draft Rectangular Counterflow Tower

The first step in the design process is to select the type of tower. Here we have selected mechanical induced-draft rectangular counterflow. The next step is to select the fill or packing type. This choice is often based on water chemistry and cleanliness of the cooling loop. Cross-fluted film fill (shown at the top of the next page) is quite popular in these towers. This packing is formed by stacking and gluing corrugated (rippled) PVC sheets at cross angles. This design provides good water and air distribution and lots of contact area to effect heat transfer. It is also fairly open so that small debris will pass through. If the water is not clean and chemical treatment does not adequately inhibit growth of unwanted organisms, fouling can occur and periodic cleaning becomes necessary.

Figure 31. Typical Cross-Fluted Film Fill

Most fill manufacturers are more than happy to provide performance curves for their products. These include mass transfer and pressure drop coefficients. We first consider the mass transfer, which is a straight line on log-log axis:

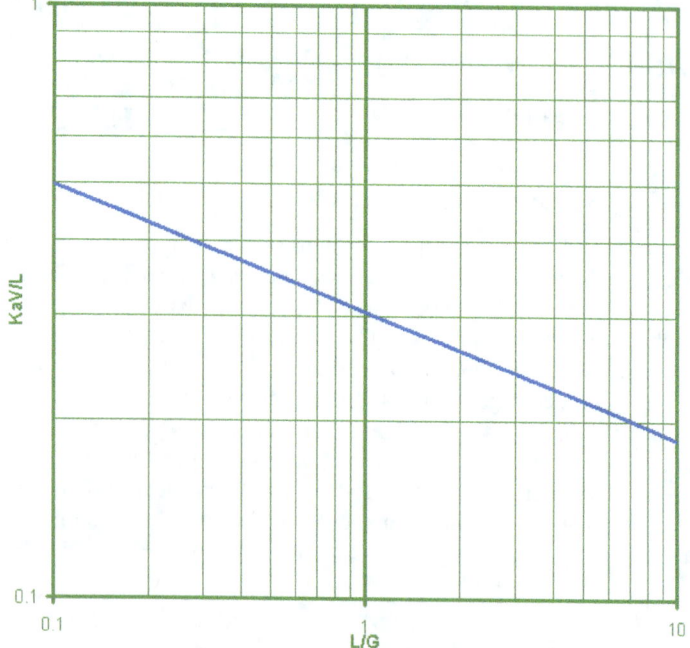

Figure 32. Counterflow Characteristic Fill Curve

Here K is the evaporative mass transfer coefficient, a is the heat exchange surface are per unit volume, V is the volume, and L is the mass flow rate of falling water per unit area. The result, KaV/L, is dimensionless. G is the dry air mass flow rate per unit area. The two areas (water and air) are equal in a counterflow arrangement. The equation of the blue curve is:

$$\frac{KaV}{L} = \frac{0.304572}{\left(\dfrac{L}{G}\right)^{0.214532}} \tag{23.1}$$

The pressure drop is shown in this next figure:

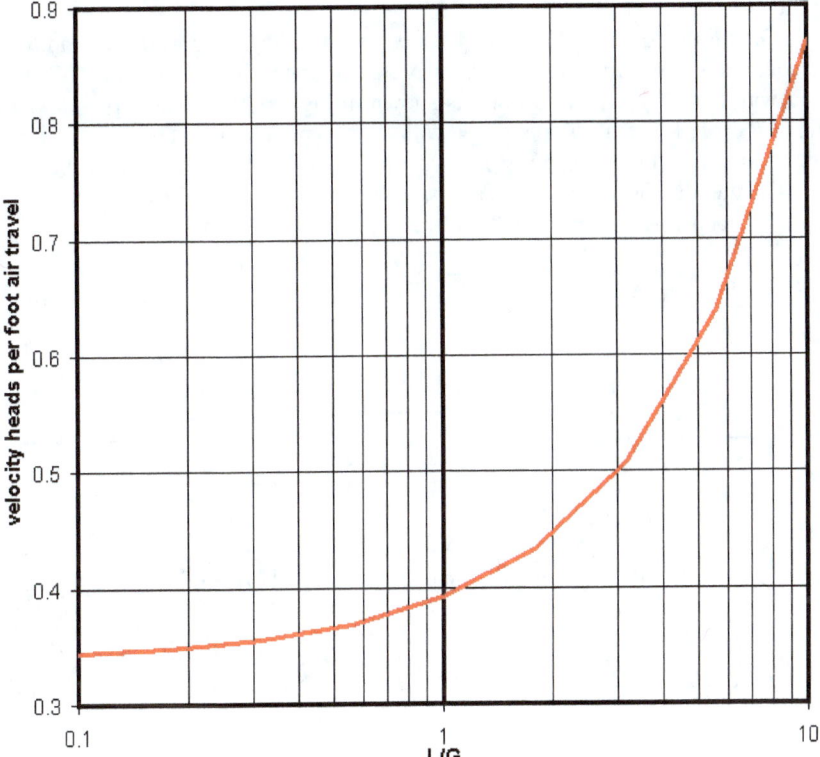

Figure 33. Counterflow Fill Pressure Drop Curve

The red curve is:

$$N_v^{'} = 0.053168\left(\frac{L}{G}\right) + 0.33833 \tag{23.2}$$

81

where N_V' is the number of velocity heads lost per foot of air travel, from which we calculate the pressure drop for the air flowing upward through the fill. This does not include any other pressure losses or gains elsewhere in the tower.

The next choice in designing such a tower is the fill (or packing) area and volume (or area and height). The area is selected so as to achieve the optimal water flux for this type of fill, which should be between 6 and 8 gpm/ft² (gallons per minute per square foot) for this type of fill. Most cooling towers are specified, designed, and tested using English units. The Cooling Technology Institute (CTI) is the world's largest organization for this industry and many members still prefers English over SI units, though arguments have been raging for many decades. I avoid such arguments and suggest that you do too, as they are pointless.

Given the desired flow of 40,000 gpm, and an approximate width to accommodate a fan at the top, we arrive at an approximate length of 140 feet and width of 48 feet with 3 fans, as shown in Figure 30. We next pick a value for L/G of 1.2 that works well with this type of fill. These values are approximate and must be refined as we continue the calculations. The fill comes in discrete depths, though it can be fabricated on site of any size for an additional cost. A common depth for this type of fill is 6 feet.

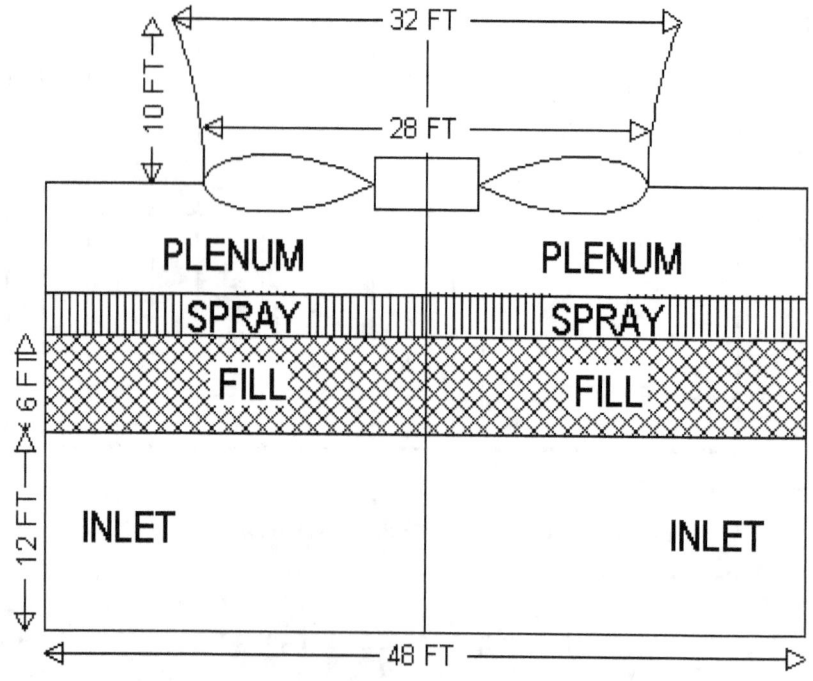

Figure 34. Cross-Section of Counterflow Cooling Tower

We must also allow for the air to enter beneath fill and 12 feet works well for small mechanical draft towers. The water is sprayed over the fill and so there must also be room for this, which depends on the spray system design. We will use 2 feet. Typical fans for this size tower are from 100 to 200 hp and 28 foot in diameter. The velocity recovery stack follows from the fan selection. There are often louvers at the inlet and these have some pressure drop as the air flows through them, perhaps 0.5 velocity heads.

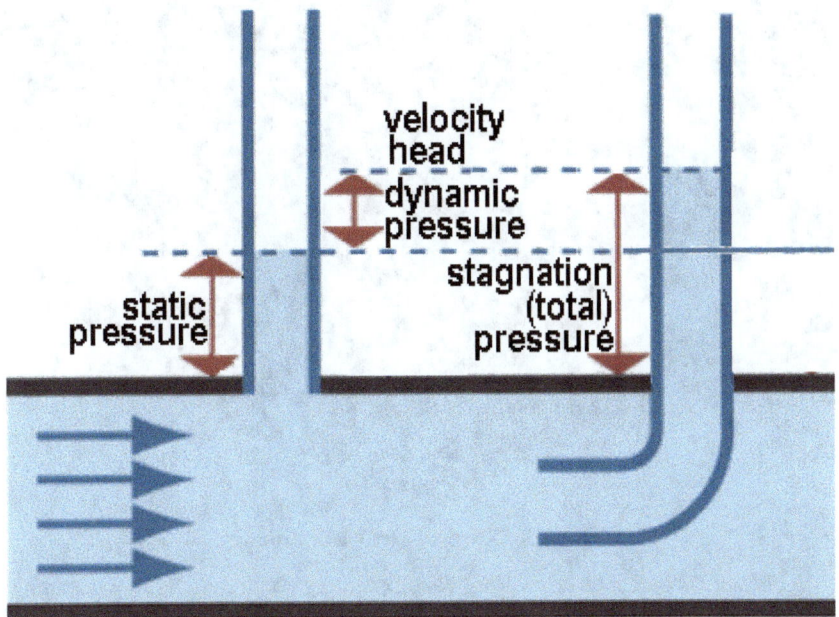

Figure 35. Static vs. Stagnation vs. Total Pressure

Much confusion exists in industry regarding pressure, what it is called, how it is measured, and how it enters into calculations. The cooling tower industry is no exception. The *static* pressure is what one typically associates with the term, *pressure*. The *dynamic* pressure (or velocity head) is associated with stagnating (halting) the flow and has a value of $\rho V^2/g$, which we know from Bernoulli's Equation (see Wikipedia for details). The *total* pressure is equal to the static pressure plus the dynamic pressure (or velocity head). In our calculations, one-half (0.5) velocity heads will be lost as the air flows through the louvers. That velocity head will be at the density and mean velocity at the face of the tower.

The velocity head will change through the tower because the density and velocity change as the air passes through and out. There will be a pressure drop as the air turns to flow upward due to the turning and that heavy rain is falling beneath the fill. There will be a pressure drop as the air flows upward through the spray. Drift eliminators are positioned above the spray in order to minimize the carry over of water droplets. Not only does this reduce water loss, it can reduce maintenance costs as well. Typical drift eliminators are shown in this next figure.

Figure 36. Typical Drift Eliminators

Pressure drop depends on design and operating conditions. For this example we will assume 1 velocity head. Note that the air velocity will be somewhat higher here than at the bottom of the fill. The *plenum* is both a contraction and shape change—from rectangular to round—directing the air into the fan.

Fan Curves

There is perhaps no greater area of confusion in industry than over fan performance curves. Many designs have failed guarantees due to this confusion. Companies have been bankrupted and lawsuits abound by misapplying fan curves. Sadly, this confusion arises from misunderstanding the pressures illustrated in Figure 35 and have been propagated for decades by industry leaders and so-called "experts" who should know better.

84

To introduce this topic, we begin with typical fan performance curves, which you might get from any one of the many manufacturers.

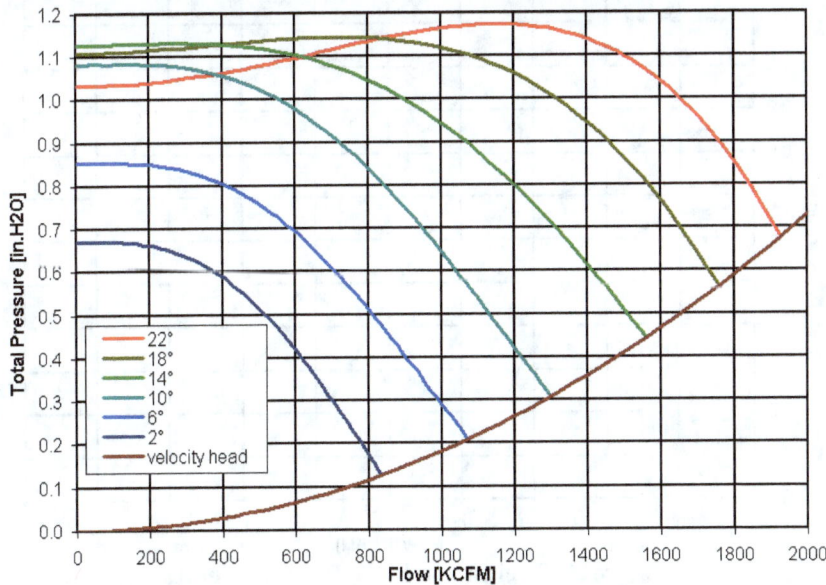

Figure 37. Typical Fan Total Pressure Curves

The dark blue, light blue, cyan, green, dark yellow, and red curves correspond to blade pitch angles of 2°, 6°, 10°, 14°, 18°, and 22°, respectively. The horizontal X-axis is volumetric flow in thousands of cubic feet per minute. The vertical Y-axis is the total (static + velocity) pressure in inches of water (one inch of water is equal to 0.24884 kPa or one kPa is equal to roughly 4 inches of water). The brown curve at the bottom is the velocity head ($\rho V^2/g$) at that flow rate. This graph is fine for illustration but it is misleading and, in fact, is the means by which the confusion began and has persisted. Why? Because we couldn't care less what the *total* pressure is across a fan. We only care about the *static* pressure. In fact, the static pressure difference times the area of the fan will equal the lateral thrust on the shaft. While we must pay for both the *static* and *velocity* pressure in the sense of supplying power to a motor, only the *static* pressure works to drive the airflow through the cooling tower or other device equipped with a fan. The *total* pressure is a useless and misleading quantity—don't use it!

We next consider the static pressure for this fan:

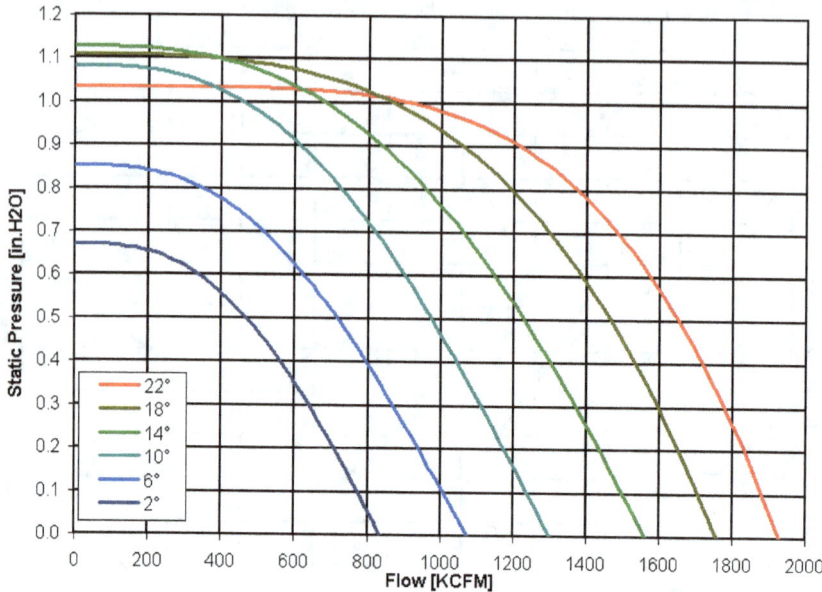

Figure 38. Typical Fan Static Pressure Curves

These curves are rarely provided by manufacturers but can be created from the total pressure curves by subtracting the velocity head (brown curve). These typically exhibit a maximum at 0 (no flow of air). Each of these curves must fall to zero at some higher flow. At this point (zero static pressure) there is no pressure rise across the fan and the fan is providing ZERO NET POWER. All of the power required to rotate the fan at this point is wasted as turbulence and friction heating. The efficiency of the fan at this point is ZERO! More importantly, you cannot get a fan to this point in a test facility (wind tunnel) without a second fan because you still must suck air into the tunnel and blow it out the other end. Here is where the money is lost: There is only one fan in a cooling tower, not like the two fans in a test facility. This operating point is completely irrelevant when designing a cooling tower or other industrial device because it will not occur in any real-world application. This only occurs in a test facility with two fans.

DON'T USE FAN TOTAL PRESSURE CURVES
TO DESIGN A COOLING TOWER!

We next consider the power required to drive (rotate) the fan:

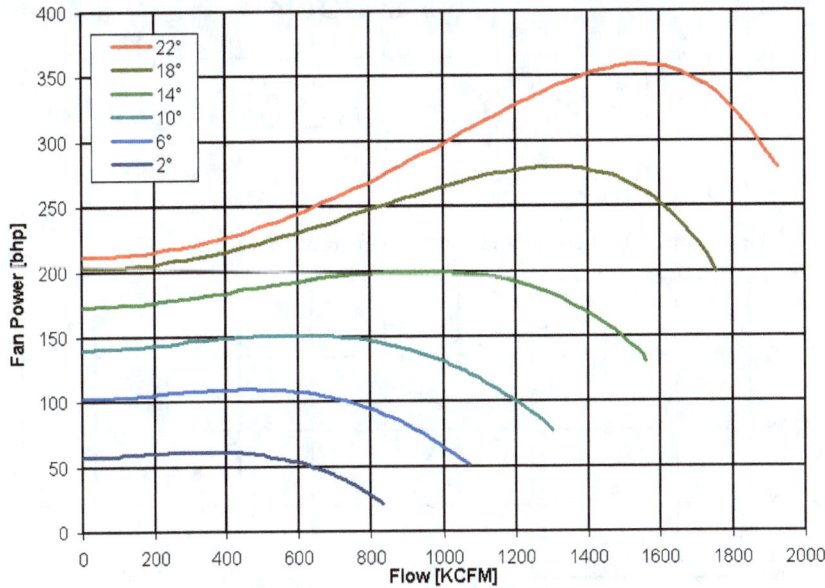

Figure 39. Typical Fan Power Curves

The hidden danger in using these curves is that some people don't know the definition of "bhp". This is *brake* horsepower, not rated or nominal horsepower. This is the power you must deliver to the shaft in order to rotate the fan. It does not include motor efficiency or gearbox loss. Those are on you. Do not neglect these in your calculations or your design will fail.

Static vs. Total Fan Efficiency

Fan efficiency is equal to the power delivered to the air divided by the power supplied by the shaft, but how is the delivered power calculated? It is equal to the volumetric flow rate times the pressure difference. As there are two pressure differences floating about (static and total), there are two efficiencies but only one is meaningful—*static* efficiency. *Total* fan efficiency is worthless hogwash. We calculate the power delivered to the air:

$$power = \dot{V}\Delta p \qquad (23.3)$$

The symbol Q is often used for volumetric flow rate and heat transfer so we make the distinction here. We must also convert units in order for Equation 23.3 to work. For example...

$$hp = \dfrac{\left(10^6 \dfrac{ft^3}{min}\right)\left(0.7\,inH2O\right)\left(144\dfrac{in^2}{ft^2}\right)}{\left(33000\dfrac{\dfrac{ft\,lbf}{min}}{hp}\right)\left(\dfrac{27.7075924\,inH2O}{\dfrac{lbf}{in^2}}\right)} = 110 \qquad (23.3)$$

The combined unit correction is approximately 1/6450.

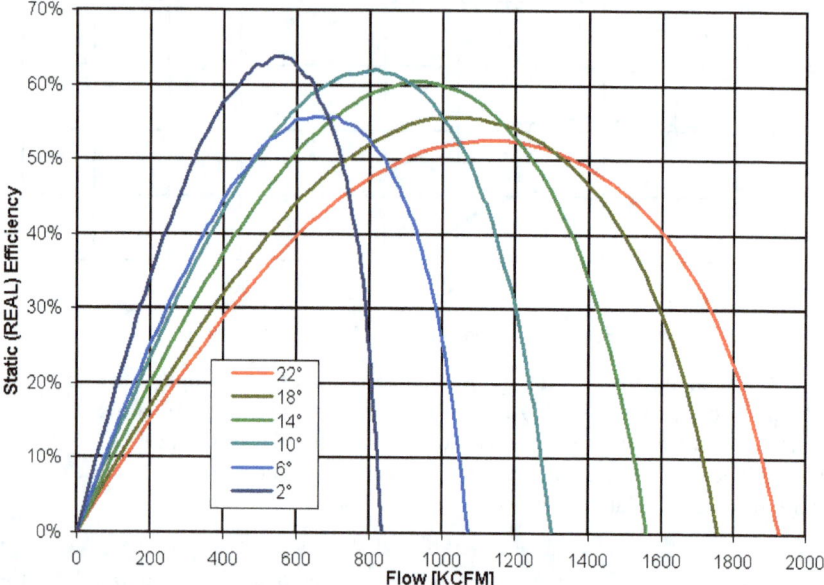

Figure 40. Fan Static (real) Efficiency

Note that the *static* (real) efficiency is zero on the left (where there is no flow) and the right (where there is no pressure rise). Remember that you can only achieve the point on the right in a wind tunnel with two fans. At that point, the other fan is pushing the air through the tunnel. This fan is just swirling air around performing no net work or power. Note also that the maximum static (real) efficiency you will ever see for an actual fan is about 70% (65% is a good fan). If you see 80%, 90%, or more, this isn't the real efficiency.

We also calculate the total (useless) fan efficiency for comparison:

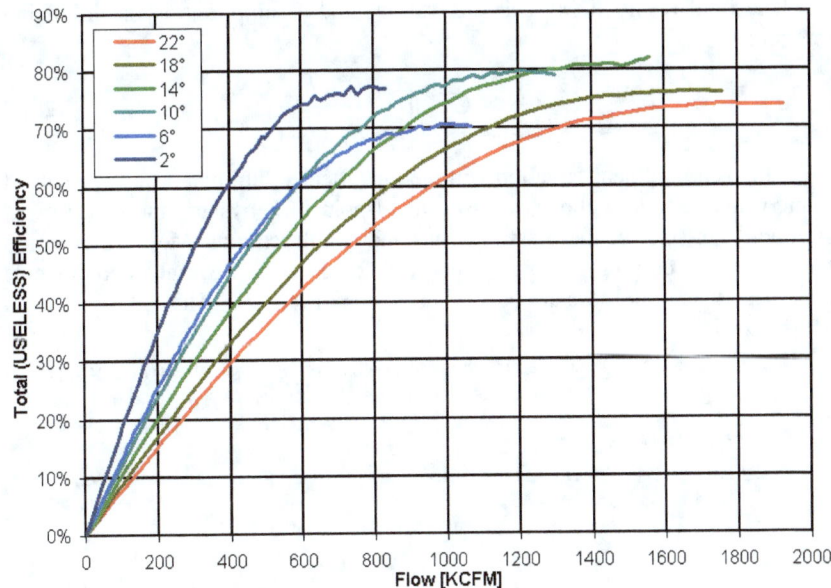

Figure 41. Typical Fan Total (useless) Efficiency

Note that the *total* (useless) efficiency, rather than dropping off to zero at the high end, plateaus. These details and graphs may be found in the online archive in spreadsheet fan.xls

DON'T EVER USE FAN TOTAL
EFFICIENCY FOR ANYTHING!

The next step involves calculating the mass transfer, *KaV/L*, which must be coupled with the pressure drop and fan curves then solved iteratively.

Merkel's Equation

In 1925 Merkel proposed a theory relating the evaporation and sensible heat transfer occurring in a direct contact process such as cooling of water or humidification of air, to an air enthalpy difference.[30] Such a representation was suited (but not limited to) various types of cooling towers. The derivation was based on counterflow contact of water and air. In fact, there were six basic assumptions that were introduced at various points in the development to simplify the mathematics. For more details on this subject, I refer you to a paper this author published with Al Feltzin.[31]

---

[30] Merkel, F. Verdunstungskuehlung, V.D.I. Forschungsarbeiten, No. 275, Berlin, 1925.

[31] Feltzin, A. E, and Benton, D. J., "A More Nearly Exact Representation of Cooling Tower Theory," CTI Technical Paper Number TP91-02.

The final result of Merkel's derivations and assumptions is the following integral:

$$\frac{KaV}{L} = C_{PW} \int_{T_{OUT}}^{T_{IN}} \frac{dT}{h_F - h_A} \qquad (23.4)$$

This is a nonlinear equation with no analytical solution and must be solved numerically. Merkel chose to use the 4-point Chebyshev method for its simplicity. In spite of these assumptions and simplifications, Merkel's Equation has served the cooling tower industry well for decades. Several improvements have been made but these are beyond the scope of this text. The VBA macro (in Merkel.xls) is:

```
Function fMerkel(baro As Double, Twb As Double, Ran As
    Double, apr As Double, rLG As Double) As Double
    Dim i As Integer, Ha As Double, Hain As Double,
        Haex As Double, Hw As Double
    Dim Tcold As Double, Thot As Double, Tw As Double,
        X(4) As Double
    X(1) = 0.1
    X(2) = 0.4
    X(3) = 0.6
    X(4) = 0.9
    Hain = fHtwb(baro, Twb)
    Haex = Hain + Ran * rLG
    Tcold = Twb + apr
    Thot = Tcold + Ran
    fMerkel = 0
    For i = 1 To 4
        Tw = Tcold + X(i) * Ran
        Hw = fHtwb(baro, Tw)
        Ha = Hain + X(i) * (Haex - Hain)
        If (Hw <= Ha) Then
            fMerkel = 999
            Exit Function
        End If
        fMerkel = fMerkel + 0.25 / (Hw - Ha)
    Next i
    fMerkel = fMerkel * Ran
End Function
```

Typical Merkel (counterflow) are shown below:

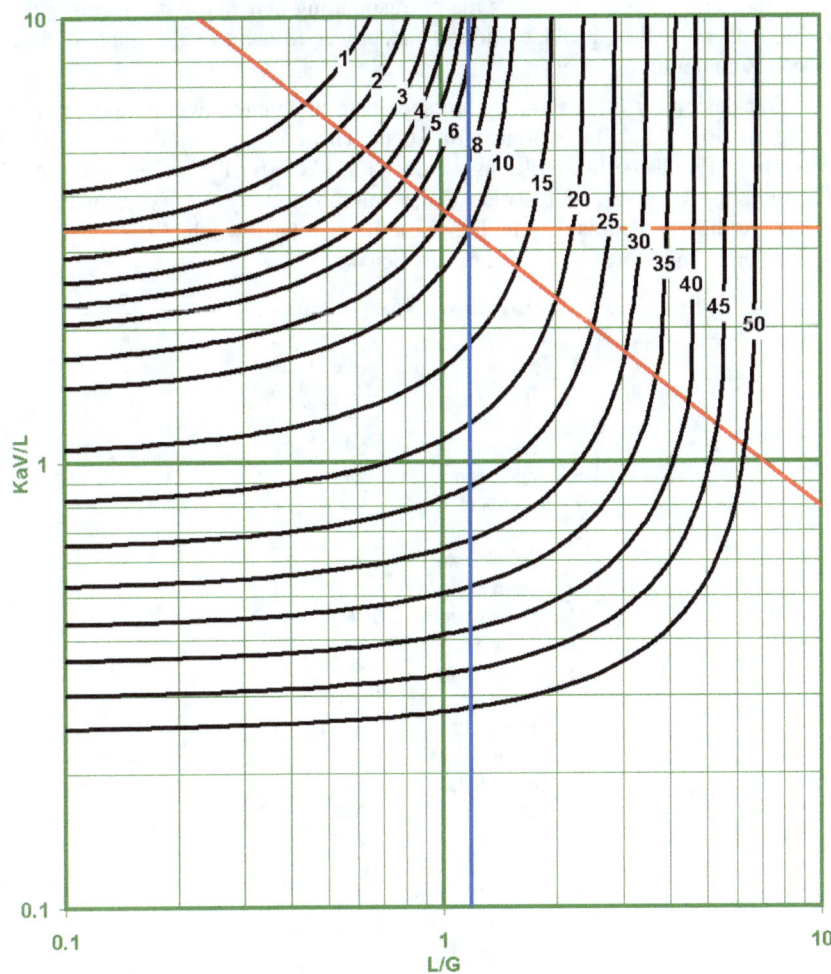

Figure 42. Typical Merkel Curves

These curves are for a barometric pressure of 14.7 psia, an ambient wet-bulb temperature of 65°F and a range of 30°F. The range is the difference in temperature of the entering hot water and the exiting cold water. The black curves are for constant approach in °F. Approach is the difference in temperature of the exiting cold water and the entering wet-bulb. The black curves are called *demand* curves. The red curve is *characteristic* of some depth of some specific fill (or packing) and may be called *supply* curves. The intersection of a black curve (desired approach) with the red curve is the operating point of the tower. Where this intersection occurs on the horizontal

X-axis is the operating *L/G* (water mass flux/dry air mass flux) shown by the blue line. The orange line shows the corresponding demand. After locating this point, you must then provide enough fan power to achieve the implied flow, which is an iterative process.

Our design for this example corresponds to an ambient wet-bulb of 65°F, a range of 30°F, and an approach of 10°F. The cold (exiting) water is then 65+30=95°F and the hot (entering) water is 95+30=105°F. The intersection of the red supply curve and the black 10°F approach curve occurs at an L/G of 1.181. Additional details may be found on the counterflow tab of the Merkel.xls spreadsheet, an excerpt of which appears below:

Counterflow Example Inputs

	A	B	C
2	INPUTS	value	units
3	barometric	14.7	psia
4	flow	40,000	gpm
5	range	30	°F
6	approach	10	°F
7	wet-bulb	65	°F
8	cells	3	-
9	fan diameter	28	ft
10	fan hub dia.	6	ft
11	fan efficiency	67.5%	-
12	stack dia.	32	ft
13	stack height	10	ft
14	stack eff.	25%	-
15	fill depth	6	ft
16	spray height	2	ft
17	inlet height	12	ft
18	fill K curve	0.600	0.670
19	fill Δp curve	0.053	0.338
20	louver loss	0.5	-
21	spray K curve	0.015	-
22	spray Δp curv	0.162	0.401
23	drift elim. loss	1.0	-

In addition to the design ambient conditions, we must define the number of cells, fan diameter, fan *static* efficiency, stack diameter, stack height, stack velocity recovery efficiency (ask the manufacturer), fill depth, spray height (above the fill), inlet height, fill characteristic curve coefficients (in KaV/L per foot plus dimensionless exponent), fill pressure drop coefficients (in velocity heads per foot of air travel in the form of y=ax+b), the inlet louver losses in velocity heads, the spray and rain mass transfer coefficients (same units and exponent as for the fill), and the drift eliminator loss in velocity heads.

Counterflow Example Calculations

	A	B	C
24	CALCULATIONS		
25	width	46.7	ft
26	length per cell	46.7	ft
27	length	140	ft
28	inlet area	4480	ft²
29	fill area	6533	ft²
30	area per fan	587	ft²
31	area per stack	804	ft²
32	water loading	6.12	gpm/ft²
33	L/G	1.181	-
34	KaV/L rain	0.08	-
35	KaV/L fill	3.22	-
36	KaV/L spray	0.03	-
37	KaV/L total	3.33	-
38	total coefficient	3.72	
39	cold wtr. temp.	75.0	°F
40	hot wtr. temp	105.0	°F
41	exit air temp.	96.4	°F
42	air below fill	65.8	°F
43	air above fill	96.1	°F
44	basin P loss	2.01	heads
45	rain P loss	3.55	heads
46	fill P loss	2.41	heads
47	spray P loss	1.18	heads
48	plenum P loss	1.28	heads
49	flow per fan	1349	KCFM
50	fan power	122	bhp
51	fan static pres	0.38633	in.H2O

The width must be larger than the fan diameter and 5/3 is a common choice, yielding a value of 46.7 feet. The length of each cell (assuming one fan per cell) must also be larger than the fan diameter and so we use the same for this. The overall length for 3 cells is then 140 feet. From these dimensions we calculate the air inlet area at the face of the tower (including all four sides), the fill area, fan area, and stack area. The water flow divided by the fill area is called the water loading and for our example is 6.12 gpm/ft². This should be somewhere between 6 and 8 gpm/ft² for this type of fill.

When calculating the required KaV/L we simply apply Merkel's formula (Equation 23.4). The heat transfer in a counterflow tower includes the fill, spray above the fill, and the rain beneath. Here we calculate each, using the same coefficients for the rain and spray, the total spray depth but only half the rain depth, because the air flows in and turns upward so that the streamlines are not equal and 50% works fairly well. Using the combined KaV/L, we draw the red curve in Figure 42. There is a macro to calculate this iteratively solving the nonlinear equation using a bisection search:

```
Function LG(baro As Double, Twb As Double, Ran As
    Double, App As Double, C1H As Double, C2 As Double)
    As Double
    Dim iter As Integer, K1 As Double, K2 As Double,
      L1 As Double, L2 As Double
    L1 = 0.1
    L2 = 10
    For iter = 1 To 32
      LG = Sqr(L1 * L2)
      K1 = C1H / LG ^ C2
      K2 = fMerkel(baro, Twb, Ran, App, LG)
      If (K1 > K2) Then
        L1 = LG
      Else
        L2 = LG
      End If
    Next iter
End Function
```

We next calculate the cold and hot water temperatures. The exiting air wet-bulb temperature is calculated by the conservation of energy using the psychrometric functions. The temperature of the air entering and leaving the fill and spray are needed to calculate the density at these locations and so we estimate these by linearly interpolating from the inlet to the exit using the KaV/L in each zone (rain, fill, and spray).

We next calculate the number of velocity heads lost for the basin (air turning upward), rain, fill, spray, and plenum. The air turns upward in the basin and the area also changes. Here we use a simple relationship for pressure drop in a pipe fitting:

$$N_{ba\sin} = \frac{5}{3}\sqrt{\frac{A_{\max}}{A_{\min}}} \qquad (23.5)$$

where N is the number of velocity heads lost, A_{min} (A_{max}) is the minimum (maximum) of the inlet and fill areas. The plenum loss is a contraction from rectangular to circular and another empirical pipe loss formula is used:

$$N_{plenum} = \left(\frac{A_{fill}}{A_{fan}} - 1\right)^{0.25} \qquad (23.6)$$

The flow per fan is defined by the number of fans, the water flow, and the L/G along with unit conversions. A constant fan static efficiency is used here. If a curve is needed, that would require iterative solution and Excel's Solver() would be a good choice to implement such a calculation. The *static* pressure rise across the fan must account for all the losses. The fan shaft power is then calculated using Equation 23.3 along with the appropriate unit conversions:

$$bhp = \frac{\Delta p \times 550 \times 60 \times 12}{N_{fans} \times 62.4 \times CFM} \qquad (23.7)$$

where Δp is in inches of water, hp = 550 ft-lbf/sec, min = 60 sec, foot = 12 inches, density of water is 62.4 lbm/ft^3, and the volumetric flow rate is in cubic feet per minute. We next consider each of these locations along the air path.

Air Path Details

	D	E	F	G	H	I	J	K	L
2		T	Pgauge	W	ρ	Area	V	head	dP
3	location	°F	in.H2O	-	lbm/ft³	ft²	ft/min	in.H2O	in.H2O
4	ambient	65.0	0	0.01327	0.07503	∞	0	0	-
5	at face	65.0	-0.04390	0.01327	0.07503	4480	840	0.04390	-0.04390
6	inside louvers	65.0	-0.06585	0.01327	0.07502	4480	840	0.04393	-0.02195
7	below fill	65.8	-0.31041	0.01364	0.07485	6533	577	0.02070	-0.24456
8	above fill	96.1	-0.36023	0.03812	0.06979	6533	619	0.02221	-0.04983
9	above spray	96.4	-0.38654	0.03844	0.06974	6533	619	0.02222	-0.02631
10	above drifts	96.4	-0.40876	0.03844	0.06973	6533	619	0.02222	-0.02222
11	fan in	96.4	-0.43727	0.03844	0.06973	1762	2296	0.30539	-0.02850
12	fan out	96.4	-0.05094	0.03840	0.06973	1762	2296	0.30539	0.38633
13	stack out	96.4	0	0.03840	0.06980	2413	1676	0.16277	0.05094

Air comes from the ambient, far from the tower at zero velocity and an infinite area. The gauge pressure of the ambient is zero by definition. This is the first position and the first line in the table above. The next location is at the face of the tower before the louvers. Air does not jump by itself from the great outdoors into your cooling tower. It takes one velocity head to suck the air out of the ambient and up to the face of your tower. Don't neglect this loss.

There is a pressure drop as the air flows through the inlet louvers, which is calculated based on the inputs. Next, the air turns and flows upward through the rain to the bottom of the fill and there are two components of pressure drop associated with this (the turning, Equation 23.5, and the rain). The air then flows upward through the fill, accumulating another pressure drop. Above the fill, the air flows through the spray with that pressure drop and then through the drift eliminators. The empirical formula (Equation 23.6) accounts for the contraction as the air approaches the fan.

Note that all of the pressure changes up until this point are negative (i.e., pressure drops). There is a rise through the fan and also a slight rise up the stack, due to recovery (slowing down in the flare) and buoyancy. Putting all these details together, we arrive at a required fan shaft power of 122 bhp. To this we must add gearbox and motor losses in order to complete the design.

Note that the gauge pressure at the stack exit is also zero, as this is open to the ambient. I have on more than one occasion measured the pressure across the exit of a cooling tower so as to experimentally confirm this boundary condition. I personally measured the pressure across the top of the natural draft cooling tower at Plant Scherer using multiple instruments.

Figure 43. Georgia Power's Plant Scherer

Example 24. Mechanical Draft Crossflow Cooling Tower

The arrangement of a mechanical induced-draft crossflow cooling tower is shown in this next figure:

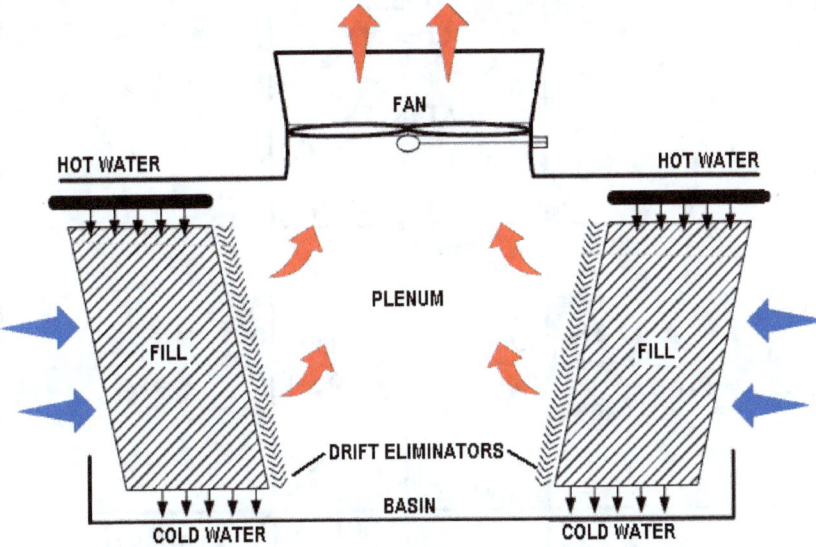

Figure 44. Typical Mechanical Draft Crossflow Tower

The water cascades vertically downward over the fill (or packing) as the air flows horizontally inward. After passing through the drift eliminators, the air turns in the plenum, passes through the fan, and out the stack. Cold water collects in the basin. The fill (or packing) is most often quite different from that used in a counterflow tower, though some fills are used in both types. The heat and mass transfer process does not occur in one dimension or counter current, rather two dimensions in cross current. Crossflow cooling tower calculations are similar to counterflow except that the mass transfer process and calculation of KaV/L is no longer a simple one-dimensional integral. We must not ignore this distinction, as some have to their ruin.

DO NOT USE COUNTERFLOW DEMAND CURVES TO DESIGN A CROSSFLOW TOWER!

While nonlinear differential equations are beyond the scope of this text, they are unavoidable in solving this problem. For more details on the subject, I suggest another text.[32] Before we consider how to solve such a problem, we consider the governing equations, including: conservation of mass, conservation of energy, Fick's Law of Diffusion, Fourier's Law of Conduction.

---

[32] Benton, D. J., *Differential Equations: Numerical Methods for Solving*, ISBN-9781983004162, Amazon, 2018.

A differential element might look like:

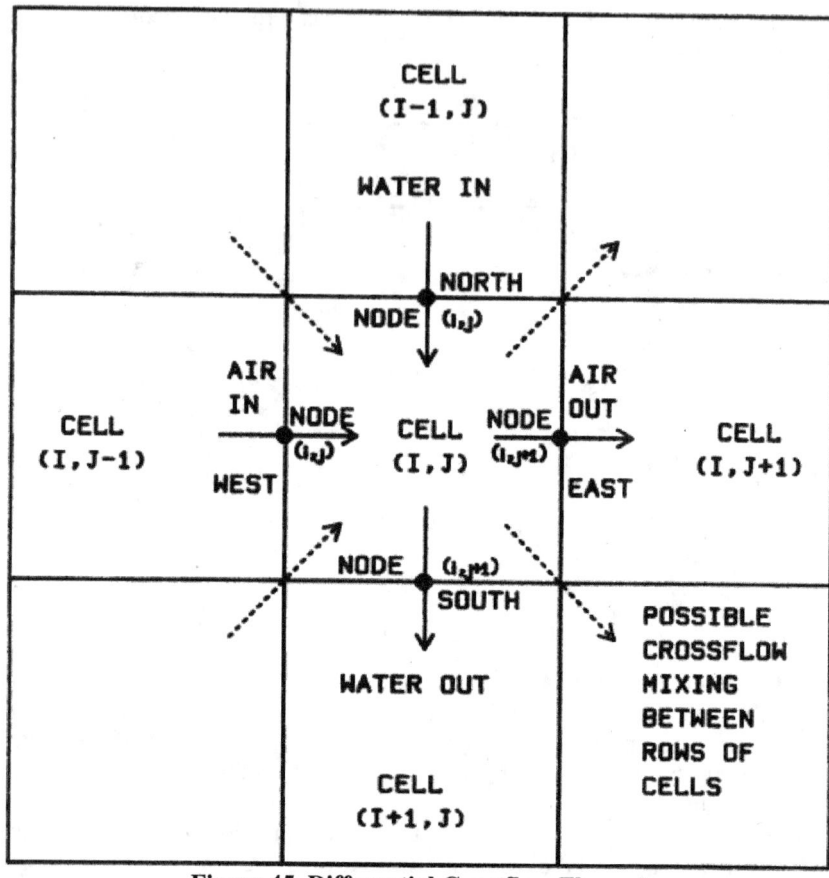

Figure 45. Differential Crossflow Element

The differential equations for a single element can be written:

$$Q_{cell} = KaY\left(h_{water} - h_{air}\right)$$
$$\frac{dh_{air}}{dA} = Q_{cell}\frac{L''}{G''} \tag{24.1}$$
$$\frac{dT_{water}}{dA} = -\frac{Q_{cell}}{C_{water}}$$

where Q_{cell} is the differential heat transfer for this cell, KaY is the differential mass transfer coefficient times height (water travel distance) for this element, h_{water} is the enthalpy of air at the entering water temperature, h_{air} is the enthalpy of air at the entering air temperature, dh_{air}/dA is differential increase in air enthalpy for the differential surface area in this cell, L''/G'' is the ratio of the

98

water mass flux (lbm/hr/ft²) to **dry** air flux (also lbm/hr/ft²), and dT_{water}/dA is the differential decrease in water temperature for this element.

Note: C_{water} is the specific heat of water equal to 1 BTU/lbm/°F, and is often omitted from this equation making the units appear not to cancel out. Do not make the mistake of converting this differential equation from English to SI units without adding the appropriate conversion factor, which is: 4.1868.

Note: The enthalpy of moist air is always per unit mass of dry air, not total mass. This distinction is unique to psychrometrics. It works in this differential equation because...

Note: G is the mass flow rate of **dry** air, not moist air. G'' is the mass flux of **dry** air, not moist air. We must be consistent with this because Geoff & Gratch developed the moist air properties on this basis. They did so for a reason: The humidity chart (see Wikipedia) allows one to solve HVAC problems graphically, which was quite important before the days of ubiquitous computers.

Note: One of Merkel's assumptions was that the combined sensible heat transfer and latent heat (mass) transfer could be approximated by the difference in saturated moist air enthalpy at the water surface (of a droplet) and saturated moist air enthalpy of the surrounding air ($h_{water}-h_{air}$) times the mass transfer coefficient, K.

Solution of this differential equation is not practical with an Excel macro. We must use a more advanced method, such as finite difference (e.g., Zivi and Brand[33] and also Kelley[34]) or Runge-Kutta (e.g., Bourillot[35]), or finite integral method (e.g., this author[36,37,38,39]).

---

[33] Zivi, S. M. and Brand, B. B., "Analysis of the Cross-Flow Cooling Tower," CTI Journal, Vol. 30, No. 1, reprinted from the Aug 1956 issue of *Refrigerating Engineering*.

[34] Kelly, *Handbook of Crossflow Cooling Tower Performance*, Neil W. Kelly & Associates, 1976.

[35] Bourillot, C., "TEFERI: Numerical Model for Calculating the Performance of an Evaporative Cooling Tower," EPRI CS-3212-SR, 1983.

[36] D. J. Benton, "A Numerical Simulation of Heat Transfer in Evaporative Cooling Towers," TVA Report No. WR28-1-900-110, September, 1983.

[37] D. J. Benton, "Development of the Finite-Integral Method," TVA Report No. WR2B-2-900-148, December, 1984.

[38] D. J. Benton, "Computer Simulation of Hybrid Fill in Crossflow Mechanical-Induced-Draft Cooling Towers," Proceedings of the ASME Winter Annual Meeting, New Orleans, Louisiana, December 9-14, 1984.

[39] D. J. Benton and W. R. Waldrop, "Computer Simulation of Transport Phenomena in Evaporative Cooling Towers," ASME Journal of Engineering for Gas Turbines and Power, Vol. 110(2), pp. 190-196, April, 1988.

First, consider a different representation of the computational cells:

						Tw				
		Ta				120.0	120.0	120.0	120.0	120.0
50	102	113	117	119	119	92.6	107.7	114.5	117.5	118.9
50	78	97	108	113	117	81.7	95.2	104.3	110.6	114.7
50	70	86	97	105	111	74.7	87.1	96.2	103.2	108.5
50	65	79	89	98	104	69.8	81.1	89.9	96.9	102.6
50	62	74	83	91	98	66.0	76.4	84.7	91.7	97.4
						Hw				
		Ha				119.7	119.7	119.7	119.7	119.7
20.3	75.2	99.7	110.7	115.7	117.9	59.6	87.2	103.7	112.2	116.3
20.3	42.0	66.9	87.2	101.0	109.4	45.6	63.7	80.1	93.9	104.2
20.3	34.3	50.5	66.8	81.8	94.2	38.4	52.1	65.3	77.7	88.9
20.3	30.3	42.3	55.0	67.5	79.3	33.9	44.9	55.8	66.4	76.6
20.3	27.8	37.2	47.5	57.9	68.2	30.8	39.9	49.1	58.3	67.3

Figure 46. Grid of Computational Cells for Crossflow

The light blue cells show air temperature. The pink cells show water temperature. The green cells show the saturated moist air enthalpy at the air temperatures in the light blue grid. The yellow cells show the saturated moist air enthalpy at the water temperatures in the pink cells. Note that there is an extra column on the left side of the light blue and green cells and an extra row at the top of the pink and yellow cells. These represent the boundary condition for the incoming air and the boundary condition for the incoming water. The average exiting (cold) water temperature is found by averaging the bottom pink row and the average exiting air temperature is found by averaging the right light blue column. Another way of looking at these cells is shown in this next figure:

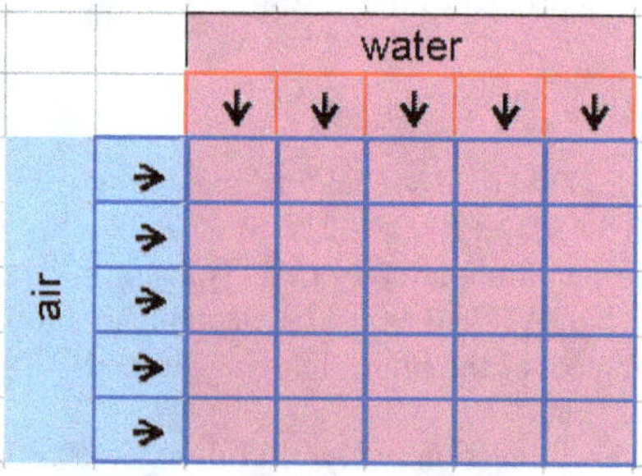

Figure 47. Crossflow Cells

The easiest way to solve this two-dimensional, two-variable, partial differential equation is using Runge-Kutta. As the inlet conditions are required for the calculations and the exit conditions are the result, we simply step through the fill horizontally, shift down one row of cells, step through horizontally again, and continue this process to the end. Equation 24.1 can be encoded in C as follows:

```
void Cell(double X,double Y,double Ha,double*dHa,double
Tw,double*dTw)
  {
  double Hw,Q;
  Hw=fHtwb(Pbaro,Tw);
  Q=KaY*(Hw-Ha);
  dHa[0]=Q*LG;
  dTw[0]=-Q;
  }
```

The process of stepping through the domain is illustrated in the following code:

```
for(y=0;y<Ny;y++)
  {
  for(x=0;x<Nx;x++)
    {
    H=Ha[(Nx+1)*y+x];
    T=Tw[Nx*y+x];
    RungeKutta2D(Cell,X,1./Nx,Y,1./Ny,&H,&T);
    Ha[(Nx+1)*y+x+1]=H;
    Tw[Nx*(y+1)+x]=T;
    }
  }
```

The Runge-Kutta code is listed in Appendix B. We can use this to create demand and supply curves similar to the counterflow ones. A Windows program to create these curves (Demand) is available free online at the location listed in the Preface. Not only will it draw the curves but it copies these onto the clipboard so that you can easily paste them into Excel or Word or some other Windows program. The numbers (digital results) can also be pasted.

We begin our example with the ambient conditions, which are a barometric pressure of 14.7 psia and a wet-bulb temperature of 65°F. The cooling range is 30°F and the approach is 10°F. Our example tower is 12 identical cells, cooling a total flow of 80,000 gpm or 6667 gpm/cell. Each cell is 32 feet long and 70 feet wide. The tower face (inward flow) area is 1920 ft² per cell and the plenum (upward flow) area is 1088 ft². The water flow area is 1152 ft².

The fill consists of plastic splash bars, as shown on the next page. This type of packing is particularly suitable for towers where the water quality is poor or occasional freezing occurs, as it is cheap and easily replaced. The fill depth (water travel distance) is 30 feet and the fill width (air travel distance) is 18 feet. These dimensions are common for this type of tower.

Figure 48. Typical Splash Bar Cooling Tower Fill

The fill mass transfer coefficient (KaY/L'' per foot of fill depth) is given by:

$$\frac{KaY}{L''} = \frac{0.06279}{\left(\dfrac{L''}{G''}\right)^{0.455}} \tag{24.2}$$

The fill pressure drop coefficient (N_V' per foot of air travel) is given by:

$$N_V' = 0.25\left(\frac{L''}{G''}\right) + 0.125 \tag{24.3}$$

Note that in both equations we use L''/G'', not L/G, as we did for the counterflow calculations. We assume 0.5 velocity heads lost across the inlet louvers and 1.5 lost passing through the drift eliminators. The fan diameter is 24 feet, hub diameter 4 feet, stack diameter 28 feet, and stack height 14 feet. The fan *static* efficiency is 65% and the stack velocity recovery efficiency is 75% (a reason for using a taller stack and smaller fan than in the counterflow example). The area per fan is 440 ft² and the area per stack is 616 ft².

Again, we must sum up all of the pressure drops from the ambient to the tower, through the louvers, through the fill, through the drift eliminators, turning and rotating in the plenum, approaching the fan, through the fan, up the fan, and out the stack. We must iteratively solve these nonlinear equations in steps, first finding the required KaY/L'' then the require fan *shaft* power, which may also include matching the fan curves so you may need to curve fit those in order to implement the complete calculation. If you leave out or overly simplify any of these steps, the design may fail. Cooling towers are very expensive and complicated. A successful design requires a team of people with various areas of expertise, as failure in any one area (thermal performance, structural integrity, water chemistry, mechanical, or electrical) can bring down the whole.

The demand curves for this condition are shown in this next figure:

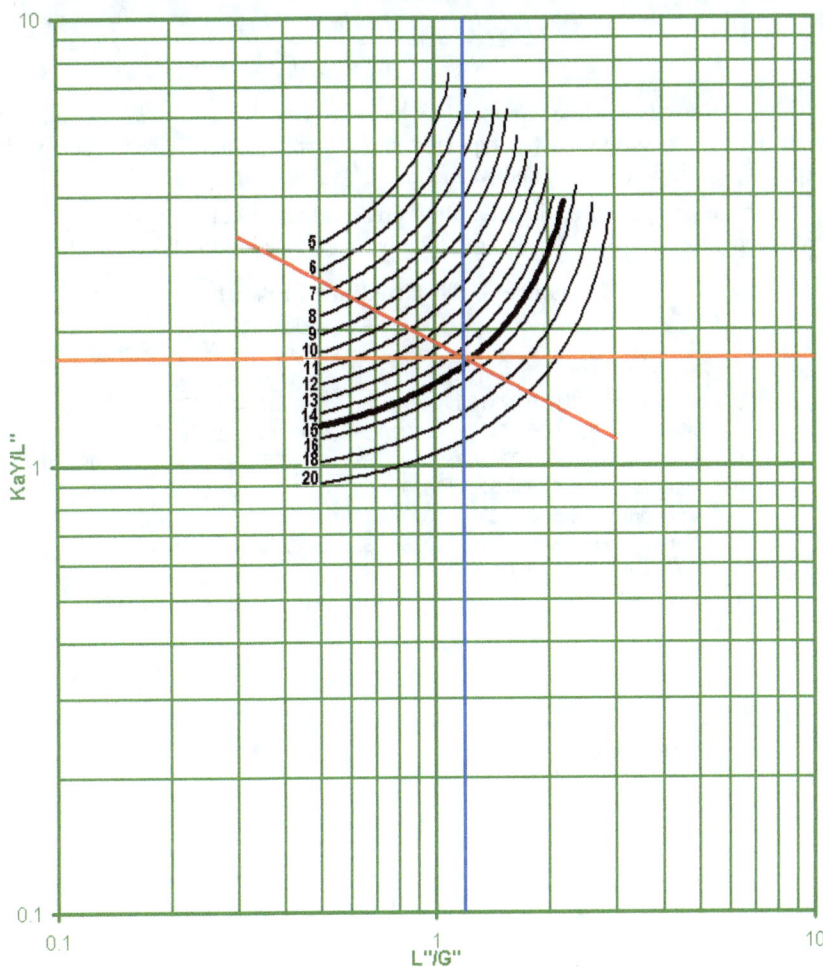

Figure 49. Crossflow Demand Curves

These were created using the free demand program with the results pasted into an Excel spreadsheet. The black curves correspond to constant approach, as listed on the left. The downward-sloping red line is the fill characteristic curve (Equation 24.2). The intercept (i.e., design point) occurs at L"/G" of 1.194, which corresponds to an L/G of 0.716, and a *KaY/L"* value of 1.673. Note that we could use *L/G* (ratio of mass flows) or *L"/G"* (ratio of mass fluxes) for counterflow, as these are the same, but we *always* use *L"/G"* for crossflow, as

103

these are not the same and this is the convention going all the way back to Zivi & Brand. The difference is the ratio of the fill depth (water travel distance) to fill width (air travel distance):

$$\frac{L''}{G''} = \left(\frac{L}{G}\right)\left(\frac{fill\ depth}{fill\ width}\right) \tag{24.4}$$

Note also that while counterflow curves typically show KaV/L on the vertical axis, crossflow curves typically show KaY/L'', the difference being fill depth, Y, vs. fill volume, V, and water flux, L'', vs. water flow, L.

The air temperature, density, flow area, velocity, velocity head, pressure drop, and gauge pressure throughout the tower are listed in the following table:

Mechanical Draft Crossflow Pressures

	A	B	C	D	E	F	G	H
1		T	ρ	A	V	head	ΔP	Pgauge
2	location	°F	lbm/ft³	ft²	ft/min	in.H2O	in.H2O	in.H2O
3	ambient	65.0	0.074108	∞	0	0	-0.01873	0
4	at face	65.0	0.074108	1920	552	0.01873	-0.01873	-0.01873
5	inside louv	65.0	0.074108	1920	552	0.01873	-0.00936	-0.02809
6	after fill	86.5	0.073261	1920	562	0.01924	-0.14667	-0.17476
7	after drifts	86.5	0.073261	1920	581	0.02006	-0.03009	-0.20485
8	before fan	86.5	0.073261	1088	2538	0.38223	-0.40018	-0.60503
9	after fan	86.5	0.073261	440	2538	0.38223	0.46340	-0.14163
10	stack exit	86.5	0.071475	616	1813	0.19501	0.14163	0.00000

Appendix A. Psychrometrics

Precise determination of the thermodynamic properties of moist air in the West began with the work of Goff & Gratch, published in a series of papers between 1943 and 1949.[40,41] This work was continued by Hyland & Wexler between the years 1978 and 1983.[42,43,44] Nelson and Sauer further refined with their research between 1999 and 2001.[45] Most recently, Hermann, Kretzschmar, and Gatley have presented a more complete formulation coupled with the latest properties of steam.[46,47]

The basic formulation has been the same since the work of Goff & Gratch. The significant peculiarity of this approach is that the properties are all on a per pound of dry air basis, rather than on a per total (air plus water vapor basis). This facilitates some calculations, which was of greater concern in the 1940s than it is now. It doesn't matter, as long as you are consistent. For the most part these properties are used for atmospheric processes, which is not a problem. As mentioned in the previous chapter and in a subsequent example, it does matter at elevated temperatures (near or above 212°F/100°C).

Saturation Pressure and the Enhancement Factor

It is tempting to simply use the saturation pressure of steam along with Dalton's Law of Partial Pressures[48] to obtain values for the water vapor content in air at saturation, but this isn't accurate. The saturation pressure of steam is for

---

[40] Goff, J. A. and Gratch, S., "Thermodynamic Properties of Moist Air," *Heating, Piping & Air Conditioning*, pp. 334-348, 1945.

[41] Goff, J. A. and Gratch, S., "Low-Pressure Properties of Water from -160 to 212 F," ASHVE Trans., pp. 95-122, 1946.

[42] Hyland, R. W., Wexler, A., and Stewart, R., "Thermodynamic Properties of Dry Air, Moist Air and Water and SI Psychrometric Charts," ASHRAE RP-216 and RP-25, 1983.

[43] Hyland, R. W. and Wexler, A., "Formulations for the Thermodynamic Properties of the Saturated Phases of H2O from 173.15 K to 473.15 K," ASHRAE Trans., Vol. 89, pp. 500-519, 1983.

[44] Hyland, R. W. and Wexler, A., "Formulations for the Thermodynamic Properties of Dry Air from 173.15 K to 473.15 K, and of Saturated Moist Air from 173.15 K to 372.15 K, at Pressures to 5 MPa," ASHRAE Trans., Vol. 89, pp. 520-535, 1983.

[45] Nelson, H. F. and Sauer, H. J., "Formulation of High-Temperature Properties for Moist Air," *HVAC&R Research* Vol. 8, pp. 311-334, 2002.

[46] Herrmann, S., Kretzschmar, H.-J., and Gatley, D. P., "Thermodynamic Properties of Real Moist Air, Dry Air, Steam, Water, and Ice," *HVAC&R Research*, 2009.

[47] Herrmann, S., Kretzschmar, H.-J., and Gatley, D. P., "Thermodynamic Properties of Real Moist Air, Dry Air, Steam, Water, and Ice - Final Report," ASHRAE RP-1485, 2009.

[48] Dalton's Law of Partial Pressures states that, in a mixture of non-reacting gases, the total pressure is equal to the sum of the partial pressures exerted by each of the constituents and these individual contributions to the whole are each proportional to the mole fraction of that component.

H2O in the vapor state in equilibrium with H2O in the liquid state. This isn't the same as H2O in the vapor state in equilibrium with air in the gaseous state. An *enhancement factor*, *f*, is introduced to account for this difference. The enhancement factor is equal to the partial pressure of water vapor that should produce the observed content divided by the saturation pressure of steam at that same temperature. The values of *f* are close to unity and the symbol is appropriate, as this is simply a *fudge factor*.

Herrmann, Kretzschmar, and Gatley, present a very complicated equation in Section 3.4.2.1 of their report for *ln(f)* in terms of second and third virial coefficients[49], explaining that this arises from Henry's Law.[50] While this is interesting and may facilitate the calculation of *f* at elevated pressures without necessitating experiments, it is immaterial. All that is needed is to measure and tabulate the water content of air. An explanation as to why it is what it is, is not essential. This is the approach that Goff & Gratch and Hyland & Wexler took. The following table of f can be found in any edition of the *ASHRAE Handbook of Fundamentals*.

Enhancement Factor, f

T°F	Pressure [in.Hg]					
	10	15	20	25	30	32
0	1.0016	1.0025	1.0033	1.0040	1.0047	1.0051
20	1.0016	1.0024	1.0032	1.0039	1.0045	1.0048
40	1.0018	1.0025	1.0032	1.0038	1.0044	1.0047
60	1.0020	1.0026	1.0033	1.0039	1.0044	1.0047
80	1.0023	1.0029	1.0036	1.0041	1.0046	1.0049
100	1.0027	1.0033	1.0040	1.0045	1.0050	1.0053
120	1.0031	1.0037	1.0044	1.0050	1.0055	1.0057
140		1.0041	1.0048	1.0054	1.0059	1.0063

The variation of *f* with temperature at 1 atm. is shown in the following. For the purpose of calculations, the humidity ratio, *W*, is needed. This can be calculated from *f*, *Psat*, and the molecular weights by Equation A.1:

$$W = \left(\frac{MW_{H2O}}{MW_{AIR}} \right) \left(\frac{fP_{SAT}}{P_{BARO} - fP_{SAT}} \right) \qquad (A.1)$$

The molecular weight of water is 18.01528 and of air is 28.9645. *Pbaro* is the barometric pressure. *Psat* is the saturation pressure of steam in the same units as the barometric pressure. The denominator in Equation A.1 becomes zero when *fPsat=Pbaro*, which is why this formulation can't be used at elevated temperatures.

---

[49] The virial expansion of the equation of state was first proposed by Kamerlingh Onnes in 1901. It forms the basis for many developments in thermodynamics related to the properties of fluids. It is... $Z=PV/RT=1+B\rho+C\rho^2+...$

[50] Henry's Law states that, at a constant temperature, the amount of a gas that will dissolve in a liquid is directly proportional to the partial pressure of that gas in equilibrium with that liquid. William Henry 1803.

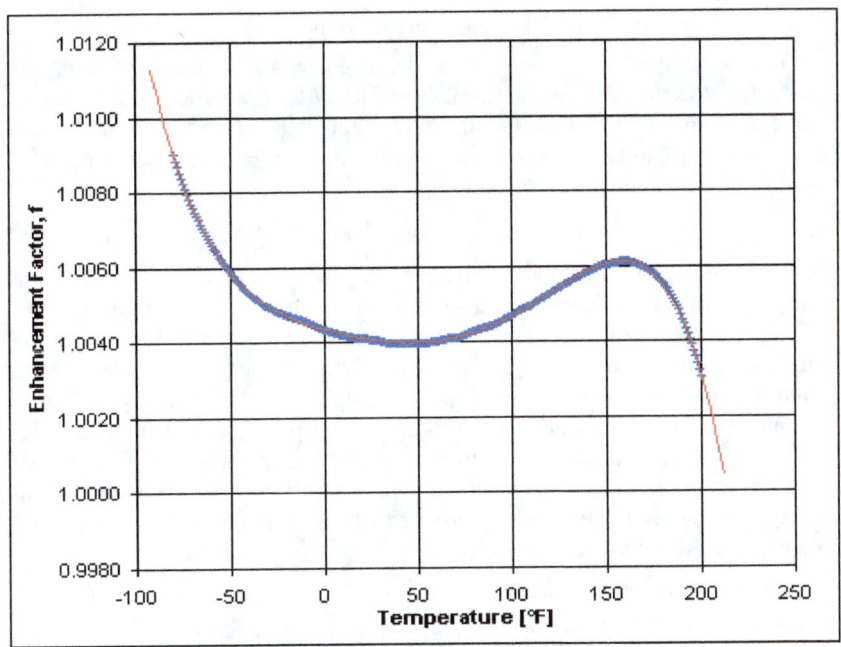

Figure 50. Enhancement Factor

The variation of *Psat* and *W* with temperature at 1 atmosphere barometric pressure is shown in this next figure:

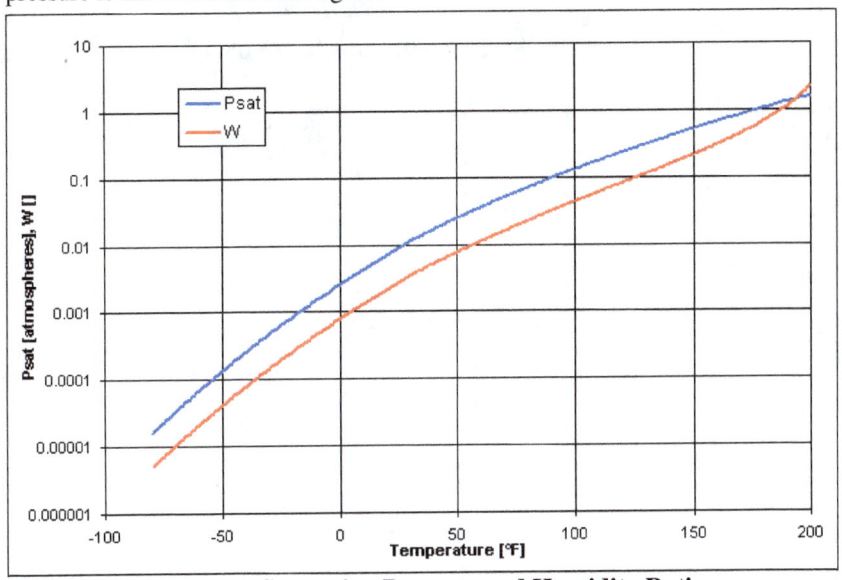

Figure 51. Saturation Pressure and Humidity Ratio

Enthalpy and Entropy

The enthalpy of moist air is also calculated on a per unit mass of dry air basis. Over the range of interest (-80°F to 212°F/-62°C to 100°C), the specific heat, Cp, of air varies so little that 0.24 BTU/lbm/°F (1 kJ/kg/°C) is an adequate representation. The enthalpy of water vapor varies linearly over this range so that the following equation is adequate:

$$h_G = 1061 + 0.444T \tag{A.2}$$

Temperature is in degrees Fahrenheit and enthalpy is in BTU/lbm. Conversion to degrees Celsius and kJ/kg is trivial. It is very important to stress here that the appropriate enthalpy of water vapor is h_G and NOT h_{FG} (that is, the enthalpy of the saturated vapor, not the latent heat of vaporization or the difference between the vapor and liquid enthalpies). Although many will insist the latter is correct-this statement even appears in print-it is not true. Consider the case of steam at the critical point flowing into the room where you now sit. At the critical point $h_{FG}=0$. You should leave immediately. You'll be thoroughly cooked long before the temperature reaches 705°F, which it most assuredly will. Energy is entering the room, but $h_{FG}=0$.

The full equation for enthalpy is:

$$h = h_A + Wh_G = 0.24\,T + W(1061 + 0.444T) \tag{A.3}$$

The entropy is a little more complicated, because of the partial pressures:

$$s_A = 0.24 \ln\left(\frac{T + 469.67}{469.67}\right) \tag{A.4}$$

$$s_G = 2.29688 - 0.00369268T + 0.0000055T^2 \tag{A.5}$$

$$s = s_A + Ws_G - R\ln\left(\frac{P}{14.696}\right) \tag{A.6}$$

A few values are listed in the following table:

Table A.1 Select Moist Air Properties

T	Ws	Ps	f	Ha	Hg	Hs	Sa	Sg	Ss
-80	0.000004948	0.000236	1.009015	-19.221	1212.6	-19.215	-0.046	-11.7	-0.046
-60	0.00002121	0.001013	1.006593	-14.414	1037.2	-14.392	-0.034	0.7	-0.034
-40	0.00007929	0.003793	1.005314	-9.609	1046.8	-9.526	-0.022	2.0	-0.022
-20	0.0002632	0.012595	1.004719	-4.804	1052.4	-4.527	-0.011	2.1	-0.010
0	0.0007875	0.037671	1.004350	0.000	1060.3	0.835	0.000	2.2	0.002
20	0.0021531	0.102798	1.004101	4.804	1069.6	7.107	0.010	2.2	0.015
40	0.005216	0.24784	1.003959	9.609	1078.2	15.233	0.020	2.159	0.031
60	0.011087	0.52193	1.004025	14.415	1087.0	26.467	0.029	2.094	0.053
80	0.022340	1.03302	1.004285	19.222	1095.7	43.701	0.039	2.035	0.084
100	0.043219	1.93492	1.004706	24.031	1104.4	71.761	0.047	1.982	0.133
120	0.081560	3.45052	1.005246	28.842	1112.9	119.612	0.056	1.933	0.213
140	0.153538	5.88945	1.005830	33.656	1121.3	205.824	0.064	1.889	0.354
160	0.29945	9.6648	1.006120	38.474	1129.6	376.737	0.072	1.848	0.625
180	0.65911	15.3097	1.005510	43.295	1137.7	793.166	0.079	1.811	1.273
200	2.30454	23.4906	1.003033	48.121	1145.6	2688.205	0.087	1.776	4.180

In this table a denotes air, g denotes gas (or vapor), and s denotes saturation. Functions to calculate the properties of moist are may be found in the on-line archive in spreadsheet Merkel.xls.

Appendix B. Runge-Kutta for Crossflow Calculations

The 4th order Runge-Kutta method for solving differential equations:

$$k_1 = f(t_n, x_n, y_n)$$

$$l_1 = g(t_n, x_n, y_n)$$

$$k_2 = f(t_n + \tfrac{1}{2}h, x_n + \tfrac{1}{2}h\,k_1, y_n + \tfrac{1}{2}h\,l_1)$$

$$l_2 = g(t_n + \tfrac{1}{2}h, x_n + \tfrac{1}{2}h\,k_1, y_n + \tfrac{1}{2}h\,l_1)$$

$$k_3 = f(t_n + \tfrac{1}{2}h, x_n + \tfrac{1}{2}h\,k_2, y_n + \tfrac{1}{2}h\,l_2)$$

$$l_3 = g(t_n + \tfrac{1}{2}h, x_n + \tfrac{1}{2}h\,k_2, y_n + \tfrac{1}{2}h\,l_2)$$

$$k_4 = f(t_n + h, x_n + h\,k_3, y_n + h\,l_3)$$

$$l_4 = f(t_n + h, x_n + h\,k_3, y_n + h\,l_3)$$

$$k = \tfrac{1}{6}(k_1 + 2k_2 + 2k_3 + k_4),$$

$$l = \tfrac{1}{6}(l_1 + 2l_2 + 2l_3 + l_4)$$

$$x_{n+1} = x_n + h\,k$$

$$y_{n+1} = y_n + h\,l$$

$$t_{n+1} = t_n + h$$

```
typedef void (*PDE)(double X,double Y,double
  U,double*dU,double V,double*dV);
void RungeKutta2D(PDE pde,double X,double dX,double
Y,double dY,double*U,double*V)
  {
  double dU,dU1,dU2,dU3,dU4,dV,dV1,dV2,dV3,dV4;
  pde(X,Y,U[0],&dU1,V[0],&dV1);
  pde(X+dX/2.,Y+dY/2,U[0]+dU1*dX/2,&dU2,
    V[0]+dV1*dY/2,&dV2);
  pde(X+dX/2.,Y+dY/2.,U[0]+dU2*dX/2.,&dU3,
    V[0]+dV2*dY/2.,&dV3);
  pde(X+dX,Y+dY,U[0]+dU3*dX,&dU4,V[0]+dV3*dY,&dV4);
  dU=(dU1+2.*dU2+2.*dU3+dU4)/6.;
  dV=(dV1+2.*dV2+2.*dV3+dV4)/6.;
  U[0]+=dU*dX;
  V[0]+=dV*dY;
  }
```

also by D. James Benton

3D Articulation: Using OpenGL, ISBN-9798596362480, Amazon, 2021 (book 3 in the 3D series).

3D Models in Motion Using OpenGL, ISBN-9798652987701, Amazon, 2020 (book 2 in the 3D series.

3D Rendering in Windows: How to display three-dimensional objects in Windows with and without OpenGL, ISBN-9781520339610, Amazon, 2016 (book 1 in the 3D series).

A Synergy of Short Stories: The whole may be greater than the sum of the parts, ISBN-9781520340319, Amazon, 2016.

Azeotropes: Behavior and Application, ISBN-9798609748997, Amazon, 2020.

bat-Elohim: Book 3 in the Little Star Trilogy, ISBN-9781686148682, Amazon, 2019.

Boilers: Performance and Testing, ISBN: 9798789062517, Amazon 2021.

Combined 3D Rendering Series: 3D Rendering in Windows®, 3D Models in Motion, and 3D Articulation, ISBN-9798484417032, Amazon, 2021.

Complex Variables: Practical Applications, ISBN-9781794250437, Amazon, 2019.

Compression & Encryption: Algorithms & Software, ISBN-9781081008826, Amazon, 2019.

Computational Fluid Dynamics: an Overview of Methods, ISBN-9781672393775, Amazon, 2019.

Computer Simulation of Power Systems: Programming Strategies and Practical Examples, ISBN-9781696218184, Amazon, 2019.

Contaminant Transport: A Numerical Approach, ISBN-9798461733216, Amazon, 2021.

CPUnleashed! Tapping Processor Speed, ISBN-9798421420361, Amazon, 2022.

Curve-Fitting: The Science and Art of Approximation, ISBN-9781520339542, Amazon, 2016.

Death by Tie: It was the best of ties. It was the worst of ties. It's what got him killed., ISBN-9798398745931, Amazon, 2023.

Differential Equations: Numerical Methods for Solving, ISBN-9781983004162, Amazon, 2018.

Equations of State: A Graphical Comparison, ISBN-9798843139520, Amazon, 2022.

Evaporative Cooling: The Science of Beating the Heat, ISBN-9781520913346, Amazon, 2017.

Forecasting: Extrapolation and Projection, ISBN-9798394019494, Amazon 2023.

Heat Engines: Thermodynamics, Cycles, & Performance Curves, ISBN-9798486886836, Amazon, 2021.

Heat Exchangers: Performance Prediction & Evaluation, ISBN-9781973589327, Amazon, 2017.

Heat Recovery Steam Generators: Thermal Design and Testing, ISBN-9781691029365, Amazon, 2019.

Heat Transfer: Heat Exchangers, Heat Recovery Steam Generators, & Cooling Towers, ISBN-9798487417831, Amazon, 2021.

The Kick-Start Murders: Visualize revenge, ISBN-9798759083375, Amazon, 2021.

Jamie2: Innocence is easily lost and cannot be restored, ISBN-9781520339375, Amazon, 2016-18.

Kyle Cooper Mysteries: Kick Start, Monte Carlo, and Waterfront Murders, ISBN-9798829365943, Amazon, 2022.

The Last Seraph: Sequel to Little Star, ISBN-9781726802253, Amazon, 2018.

Little Star: God doesn't do things the way we expect Him to. He's better than that! ISBN-9781520338903, Amazon, 2015-17.

Living Math: Seeing mathematics in every day life (and appreciating it more too), ISBN-9781520336992, Amazon, 2016.

Lost Cause: If only history could be changed..., ISBN-9781521173770, Amazon, 2017.

Mass Transfer: Diffusion & Convection, ISBN-9798702403106, Amazon, 2021.

Mill Town Destiny: The Hand of Providence brought them together to rescue the mill, the town, and each other, ISBN-9781520864679, Amazon, 2017.

Monte Carlo Murders: Who Killed Who and Why, ISBN-9798829341848, Amazon, 2022.

Monte Carlo Simulation: The Art of Random Process Characterization, ISBN-9781980577874, Amazon, 2018.

Nonlinear Equations: Numerical Methods for Solving, ISBN-9781717767318, Amazon, 2018.

Numerical Calculus: Differentiation and Integration, ISBN-9781980680901, Amazon, 2018.

Numerical Methods: Nonlinear Equations, Numerical Calculus, & Differential Equations, ISBN-9798486246845, Amazon, 2021.

Orthogonal Functions: The Many Uses of, ISBN-9781719876162, Amazon, 2018.

Overwhelming Evidence: A Pilgrimage, ISBN-9798515642211, Amazon, 2021.

Particle Tracking: Computational Strategies and Diverse Examples, ISBN-9781692512651, Amazon, 2019.

Plumes: Delineation & Transport, ISBN-9781702292771, Amazon, 2019.

Power Plant Performance Curves: for Testing and Dispatch, ISBN-9798640192698, Amazon, 2020.

Practical Linear Algebra: Principles & Software, ISBN-9798860910584, Amazon, 2023.

Props, Fans, & Pumps: Design & Performance, ISBN-9798645391195, Amazon, 2020.

Remediation: Contaminant Transport, Particle Tracking, & Plumes, ISBN-9798485651190, Amazon, 2021.

ROFL: Rolling on the Floor Laughing, ISBN-9781973300007, Amazon, 2017.

Seminole Rain: You don't choose destiny. It chooses you, ISBN-9798668502196, Amazon, 2020.

Septillionth: 1 in 10^{24}, ISBN-9798410762472, Amazon, 2022.

Software Development: Targeted Applications, ISBN-9798850653989, Amazon, 2023.

Software Recipes: Proven Tools, ISBN-9798815229556, Amazon, 2022.

Steam 2020: to 150 GPa and 6000 K, ISBN-9798634643830, Amazon, 2020.

Thermochemical Reactions: Numerical Solutions, ISBN-9781073417872, Amazon, 2019.

Thermodynamic and Transport Properties of Fluids, ISBN-9781092120845, Amazon, 2019.

Thermodynamic Cycles: Effective Modeling Strategies for Software Development, ISBN-9781070934372, Amazon, 2019.

Thermodynamics - Theory & Practice: The science of energy and power, ISBN-9781520339795, Amazon, 2016.

Version-Independent Programming: Code Development Guidelines for the Windows® Operating System, ISBN-9781520339146, Amazon, 2016.

The Waterfront Murders: As you sow, so shall you reap, ISBN-9798611314500, Amazon, 2020.

Weather Data: Where To Get It and How To Process It, ISBN-9798868037894, Amazon, 2023.